AI TECHNOLOGY BRINGS POSITIVE OR NEGATIVE TO

IMPACT OUR SOCIETIES

JOHN LOK

Table of contents

and productivity
5.5 What occupations will be influenced by (AI) technology.

What does artificial intelligence(AI) mean? p.87-102

● What (AI) function is?
● Can (AI) impact human job nature?
● How can human society job nature
to be changed to artificial intelligent society?
● Why does human need artificial intelligence machines?
● How does artificial intelligence influence future working
changing in automation employment and productivity aspects?
● Is artificial intelligence possible to replace labor ?
● Can (AI) technology replace human labour nature of work?
● Why can artificial intelligence satisfy human needs?
● Is artificial intelligence one good choice for human future
technological benefit?

What is relationship between
(AI) and economy growth? p.104-162

● How can artificial intelligence technology influence economy?
● Can (AI) influence global economy growth?
● How can artificial intelligence impact global economy growth?
● What is the relationship between (AI) and (CRM)?
● What is the relationship between
(AI) and global digital economy development ?

● How can (AI) technology influence digital economy?
● Could work activities in China be automated
making in the nation with the world's largest automation
potential?
● Will (AI) technology influence digital economy change to
manufacturing industry ?
● What is artificial intelligence potential

16.4 Why model of sustainable development environment industry be (AI) robot attractive market
16.4.1 How to invent (AI) brain's environment protection ability?

16.5 (AI) Asia market

16.6 Is the brain a good model
for machine intelligence?
Reference p.356

Does need to keep (AI) legal

4.1 Ethical and theological reflections
on artificial intelligence
4.2 Social choice ethics in artificial
intelligence p.357-376

- (AI) weapon moral or immoral
invention

Artificial intelligence and the future of defense

(AI) system immoral intention

(AI) soldier weapon ethical, social and
economic negative impact

Reference
Behavioral ethic in business
environment
1.1 Business ethic case
How scientifical ethic influences

Contents

PREFACE

Introduction

What is future (AI) artificial intelligent products development trend and reasonable development stages? How to predict consumer behaviors to persuade who to feel (AI) products are more satisfactory to their needs? Why do consumers feel them to need to buy any (AI) products to use? Will it have other similar products to replace (AI) any products? How is the reasonable stages to achieve future (AI) development in success? Can AI development bring global economic growth or avoid recession? Can future non-manual driving public transport tool inventions bring global economic growth? AI can bring positive or negative to impact our society.

In this book, I shall give actual data to predict what the future (AI) products development trend is. Giving my opinions to predict how (AI) consumers' choices are more absolutely. In the (AI) past first stage, I shall concern travel, education, transportation, financial , hospital, administrative service etc. different job natures to indicate how to apply (AI) products to assist these industries more beneficial. In the (AI) nowadays second stage, I shall concentrate on how (AI) developing on education aspect. In the (AI) future third stage, I shall explain why (AI) will have possible to invent (AI) brain technology, even it will bring (AI) war occurence in possible.

In the first (AI) stage, it concerns to be given my opinions to explain how artificial intelligent technology will impact our life and will influence economic development in the future as well as how to influence human job market change. In (AI) labor market stage, I shall indicate how artificial intelligence technology influences future macro global economy change.

Advances in artificial intelligence (AI) technology is for the progress in critical areas, such as health, education, energy, economy inclusion, social welfare and the environment. Thus, it brings this question: Which (AI) workers be instead of traditional human workers in these different new markets? In recent years, machines had been used to be human's tasks in the performance of certain tasks related to intelligence , such as aspects of image recognition. Experts also forecast that rapid progress in the field of specialized artificial intelligence will continue. Then, it also brings this question: Does (AI) exceed that of human performance on more and more tasks? If it is truth, will some of human jobs to be disappeared? (AI) will be instead of human some simple jobs, then unemployment rate to the low skillful and low educated workers will be increased.

Whether (AI) will be raised either production or performance or unemployment to bring human job market more advantages or more disadvantages? In my this book, I shall explain whether (AI) will bring benefits or disadvantages to human job market. I shall give example to let my readers to think how to support my final view point.

What is future (AI) artificial intelligent products development trend? How to predict consumer behaviors to persuade who to feel (AI) products are more satisfactory to their needs? Why do consumers feel them to need to buy any (AI) products to use? Will it have other similar products to replace (AI) any products?

In (AI) second stage, I shall indicate how (AI) is developed on education aspect, artificial intelligent technology and online technology and online book stores are high technological intelligent product. Hence, human ourselves will have possible to cause artificial intelligent machine men to own human's mind to learn how to read books and/or write books abilities. When artificial intelligent machine men can learn how to read books and/or write books. Consequently, it means that artificial intelligent machine men can own human mind to do any jobs.I shall assume when artificial intelligent machine men can learn how to write books and/or read books. Then, they will have human's mind ability in possible. Can future artificial intelligent machine men be invented to learn how to write books and/or read books ability? In my this book, I shall attempt to answer this question. Finally, I hope my readers can attempt to make judgement whether artificial intelligent machine men can really learn how to write and/or books. I shall apply online technology to answer this answer. Finally, I shall give my opinions what are the influences when AI machine men had invented to achieve owning human's mind and judgement ability to our future society.

In (AI) third stage development, artificial intelligence (AI) technology is popular to be applied to different industry aspects, such as medical, construction, transportation, hospital, education etc. Although, (AI) is a human invention new development. IN fact, it seems only beneficial to human's daily life. But, it will also have threats to influence human's safety in possible , if some scientists or self-interest mind people who aim to apply (AI) to earn more profit or apply (AI) tools to be weapon to attack other countries to achieve to dominate all human's ambitious intention. Thus, (AI) will bring negative influences to our society, instead of positive influences if we can not apply this kind of new technological tools immorally. In this stage, I shall give my opinions to indicate what reasons will cause (AI) artificial intelligent tools to be applied to social military defense weapon by human's intention. In my this books, I hope my readers can know what will cause human's immoral behaviors to bring our societies to bring more dangerous or risks or threats if human applied (AI) technology to achieve whose immoral or ambitious intention. Finally, I hope that human ought not apply (AI) technology to do any behavioral attack to satisfy ourselves interest or dominate global world ambition to avoid (AI) technological war occurrence in the future one day.

In (AI) fourth stage, I shall give some university lecturers' personal analytical mind to judge whether what will be occurred if artificial intelligence could be invented to match to own human brain's mind ability? What will be the advantages and/or disadvantages if (AI) robots would be invented to match to own human brain's mind ability? I shall follow current (AI) technological development to judge whether what the potential abilities are that (AI) will achieve to satisfy human's life needs when (AI) is invented to own human brain in future one day.

In AI fifth stage, we are experiencing technological development stages, human began to consider air pollution how brings our natural environment to cause worse. Our cars fuel emission can also pollute our air when we are often driving ourselves cars to go to anywhere for leisure or working aim. Instead of ourselves cars emission air pollution, public transport tools, such as buses, taxis, ferries, their fuel emission also influence our air to be polluted. Because passenger individual comfort need began raises. So, electric auto non-manual driving cars, battery charge energy cars are began to consider how to invent in order to reduce air pollution and raise driver individual comfortable need, instead of improving driving speed to be rapid, car inventors also hope to avoid air pollution to pollute our natural environment. When global fuel cars number continues increases.

I shall indicate how our future transportation tool may be invented in order to improve our qualify of lives or standard of lives to be better. How to improve our future transport tools in order to avoid air pollution more serious? It is my this book discussion about how transport invention or improvement, it will influence public transport passenger service and comfortable need aim. I shall attempt to research how transport improvement question in order to let readers feel how it can influence future public transport tools passenger individual choosing any kinds of pubic transport tools need as well as how is our future actual transport improvement achievement aim in order to keep our standard of lives (quality of lives) can be improved to achieve the best service standard and avoid any kinds of future public transport tools passengers number reduces. Because any public transport needs to be improved in order to satisfy passengers comfortable needs. IN passenger comfortable need psychology view, how to improve public transport service quality. It is my main discussion in this book topic, it concerns bus, ferry and rail public transport passenger psychology.
How man made clever automobile may well engage clever transportation tools ? Can man made intelligence (AI) and laptop

studying (ML) be utilized in the quest for brand spanking new " intake" behavioral kind variables that have an effect on purchaser person or transportation carrier association person diverse transportation instruments offerings, corresponding to street or sea or sky transportation instruments? Can man made clever car might engage clever transportation instruments industry development?

This road transport part brings readers to image what will be different if artificial intelligent non manual driving vehicle will be used to public transportation and private transportation both aspects in popular. Will it popular to accept to use any artificial intelligent vehicles? Is it possible to apply AI non-manual driving technology to AI non-manual driving transportation tools global transportation market? For example, in (AI) non-manual vehicle industry, driving automatic vehicle whether it will be accepted to drivers who have confidence to drive it on roads safely. Whether artificial (AI) intelligent non-manual driving systems are the improvement of traffic safety, reduction of energy consumption or improvement of the comfort of the driver. Whether will it be popular to accept to apply artificial intelligent non-manual driving technology from non-manual auto driving cars to be applied to any non-manual auto driving transportation tools transportation market development, such as train, tram, lorry, transportation air plane, passenger air plane, ferry, taxi, MTR. Etc. different kinds of transportation tools? If future human accepts to use any non-manual driving vehicles or non-manual driving transportation tools, what advantages and disadvantages will bring to influence our daily life.

How if (AI) non-manual auto driving technology stage is mature to achieve non-manual driving technology is safe driving. It is possible that (AI) non-manual driving cars can influence to change whole manual driving transportation tools to non-manual driving transportation tools. How it will influence (AI) autonomous cars change to influence global manual driving transportation industry development ? To achieve non-manual driving industry development success. (AI) non-manual driving vehicle manufacturers need to ensure (AI) driving system is more safe to drive to compare manual driving on the road. If they expect non-manual driving transportation market development success. So, self improving systems are a promising new approach to developing artificial intelligence. But will their behavior be predictable? Can will be sure that they will behave as we intended even after many generations of self improvement? This part presents a framework for answering any questions concern whether future non-manual driving transportation market whether it will be possible to bring global economic growth in success.

In AI final stage, I shall indicate whether AI can assist space tourism development in possible as well as what benefits it can bring to our societies. I write this book aims to let readers to feel whether AI can bring real benefits to our future societies in possible.

Prologue

Table of contents

Why and how non-manual driving car owners need raise public transport quality on travel time and fare aspects

How non human driving behavior can be influence by non-manual driving cars

Artificial Intelligent In Road Transportation Strategy

HOW DESIGNING UNDERGROUND MASS TRANSIT RAILWAY TO BRING PASSENGERS

- Designing transportation system advantages

Artificial intelligent public transport how influences passenger psychology

How technology influence passenger psychology p.447-467

CAN NON-MANUAL DRIVING AIR PLANES EXCITE FUTURE AIRLINE AND AIRPORT TRANSPORT INDUSTRY DEVELOPMENT

Will airline industry's ticket price elasticity be influenced by demand and supply factor

- Boeing 747 manufacturing fuel cost strategy
- How airlines and airports implement successful network strategies
- Performance measurement system strategy
- What factors influence cost-related management quality ? p.468-495

Chapter 5 AI sixth stage development

How AI technology assists human to avoid energy and food waste

Factors influence householder energy using behavior at home

House quality influences householder electricity energy consumption behavior p.496-519

Environmental impacts of householder greenhouse gas electricity energy

The effect of house space occupancy and building characteristics on householder electricity energy use

(AI) robotic technology brings long term or short term benefits to space exploration industry

● (AI) technological Space tourism

Future space tourism psychology prediction stragegy

● Psychology and economic environment changing both factors influence whole space tourism market leisure desire

● space tourism strategy

(1) safe space tourism journey

(2) reduction cost expense plan

(3) achieve any space tourism mission plan

Methods to raise space traveler number p.571-585

● What is the prediction space travelling passenger desire method ?

● The prediction of price factor influences space traveler number

The influential factors persuade travelers choose space tourism

Raising space tourism leisure consumption strategies

Space travel markting strategy p.586-595

(1) On concept of spacecraft design aspect

● Outsourcing spacecraft concept design strategy

(2) On deciding mission aspect

(4) On target audience prediction aspect

(3) On space tourism leisure organization management aspect

(5) On space objective aspect

Space tourism leisure behavioral economic consumption model

(1) Economic environment variable factor

(2) Space tourism leisure journey management factor

Space tourism market moral ethic risk threats

(1) Potential accidents aspect

(2) Space tourism destinations and space tourism entertainment

I

(AI) development first stage

Competitive influences between artificial intelligence and human job

Although, (AI) technology will be popular to applied to different jobs, but it still needs social acceptance to replace some human jobs. Today, it is increasingly common for people to use robots in various situations at home and in retail stores, hotels and hospitals. Robots are classified into several types based on their functionality (service and utility robots or those designed to communicate with humans) and appearance (humanoid robots or mechanical robots). The types of robot to which every country attaches particular important in the advance of robotics, reflects the sense of values and preferences of its population . Thus, (AI) will be applied to replace human to do these above different kinds of job nature. For example, U.S. has the highest level of robot utilization at home and an retail stores with its people being the most enthusiastic about the future use of robots. Otherwise, Germany shows a strong tendency to consider robots for industrial purposes, and its people feel strong to the presence of robots in their households. Japanese accepts to apply" human aid robot" that can communicate with humans and they have a high level of familiarity with robots.

Hence, it implied those three countries have accept (AI) to replace human to do any these kinds of job duty and it will influence these three countries' workers lose their old occupations and who will unemployed absolutely, due to many (AI) robots replace them to do their job duties in the future. Also, US will have many retail service workers or retail warehouse workers are unemployed. Germany will have many manufacturing industry's workers are unemployed. Japanese will have many communication industry workers are unemployed, such as telephone service, shopping center services etc. different kind of service industry's service staffs . It will cause these kind of workers' competitive abilities are lost in themselves countries' jobs that require such skills include software developers, court judges, nurses, high school teachers, dentists and university lecturers, these occupations are still difficult to be replaced by (AI) robots.

Are robots taking our jobs or making them? In fact, our societies will have unemployment challenges, even (AI) technology has not created before. However, after (AI) robots invention, some of human jobs will be replaced and it can raise many low skillful and low knowledge level worker unemployment number. However, I think that high productivity driven by increasingly powerful IT -enabled machines is the causes of global labor market problems and accelerating technological change will only make those problems worse.

IT technology brings this question: Are robots killing human's jobs or benefiting human's jobs? I suppose that there is a limited amount of labor to be done. The implication is that technology can create unemployment by displacing workers, such as (AI) invention, because the more efficiently worker work (using machines or (AI) robots), the loss work there is for workers to do. Even, any new jobs will be better done by machines or (AI) robots, and unemployment will still skyrocket. How do we know that humans will always be better at some work, or more importantly, enough work, than machines or (AI) robots, e.g. human drivers drive more safe or careful to compare (AI) robot drivers. But, the challenge is that it is not ensure that (AI) robots drivers must not drive careless to cause the chance of accident occurrences more than human drivers. However, technological change can be beneficial to innovation, automation and increasing productivity for businesses.

Consequently , it may seem machines can hurt wages and job for low skillful, less educated workers. Also, high educated workers are likely as less educated workers to find themselves displaced and devalued, and more education may create as many problems as it solves. Thus, in negative influence, automation effects on particular jobs shift workers to other jobs that are equally or more desirable. Workers may be highly compensated for possessing human capital that is specialized to a labor market. If technological advance is very rapid, such as (AI) invention, causing a large and very rapid drop in demand in a large labor market, the economy may not be able to absorb the sudden surplus of labor in a short period of timer when (AI) robots are popular to replace some workers to do some occupations in global societies.

For example, self-driving vehicles threaten to send truck drivers to the unemployment office. Computer programs can now write journalistic accounts of sporting events and stock price movement. There are even computers that can grade essay revolutionize some part of teaching jobs. Hence, (AI) robots will have possible to replace human brain to do any judgement, argument, and mind job duties. It implies some occupations which need human' mind will be threaten by (AI) robots, e.g. author, accountant, nurse, engineer. Thus, (AI) robots will have possible to replace some professional and high educated workers' jobs in the future.

But, technology can create new nature of jobs in possible. For example, a 60 minutes program indicated technology is putting new categories of jobs in the sites (sic) of automation, the 60% of the workforce that makes its living gathering and analyzing information. Also, recession: technology kills middle -class jobs that overall technology is eliminating for more jobs than it is creating by (AI) technology. Hence, human's brain work may be assisted by 60%

of (AI) gathering and analyzing information for some occupation , e.g. space scientists, ocean scientists, earth scientists etc.

However, I believe the (AI) invention and human job competition may influence global productivity change. Productivity is economic output per unit of input, the unit of on input can be labor hours(labor productivity), but if (AI) robots replace human job, then the unit of input may be (AI) machine hours (AI) robot productivity or all production factors including labors, machines and energy (total factor of productivity). Producing more output with less input can take several forms.

The traditional notion of productivity is a form reorganizing production and/or using better or more technology to produce more output per worker hour. But when (AI) robots invention, the form can be reorganizing production and/or using better or more (AI) robots to produce more output per (AI) robot hour. Hence, if the firm apply (AI) robots to produce its products. Then , productivity improvements in the firm may result in less workers employment, due to (AI) robots replace more worker number to achieve more productivity improvement, it has economic benefits (less factor of production) , but more production in long term.

Thus, (AI) robots can help any firm to achieve productivity improvement in long term, for example, if unproductve farmers move to the city and start working for high-tech. manufacturers. The shift effect can be more dynamic and disruptive as low-productivity industries lose out in the marketplace to high -productivity industries and the compositional mix of the economy changes. Thus, in the long term (AI) robots can also be beneficial to high productivity industries to bring the mix of economy positive changes.

Moreover, automation will also produce some new jobs in firms that sell the new robot or other labor-saving technology. This means that, in general, there will be shift in the economy in the direction of higher-skill and higher wage jobs. Even if the (AI) robot invention country, US becomes a leader in (AI) robots producing productivity-enhancing technology, it will experience a growth in jobs serving foreign (AI) robots product buyers. Hence, (AI) robots can also create (AI) salespeople, (AI) manufacturing workers , (AI) inventors, scientists, (AI) software designer etc. occupations, when if all society does is move workers from insurance firms, restaurants and car factories to robot factories, productivity will have remained the same to create job needs for insurance, restaurant and car manufacturing worker service occupations for (AI) software designer, (AI) service robots manufacturer, (AI) service robot seller etc. related (AI) service robot product occupation created in (AI) robot technology job market. Hence, (AI) invention also create new (AI) technology job chance. (AI) impacts management job market.

In future, organization management will be changed from (AI) introduction. Division of labor will change and collaboration among humans and machines will increase. Companies will have to adapt their training, performance and talent acquisition strategies to account for a new found emphasis on work that hinges on human
judgement and skills, including experimentation and colloboration.

How (AI) impacts any organizational administrative management work? (AI) 's greatest impact will be on administrative coordination and control tasks, such as scheduling, resource allocation and reporting, (AI)-driven will place a higher premium on what we call " judgement work", the application of human experience and expertise to critical business decisions and practices when information available is insufficient to suggest a successful course of action. This kind of work will require new skills and mindsets; replacing people with machines is not goal in itself. When, artificial intelligence enables cost-cutting automation of routine work, it also empowers value -adding augmentation of human capabilities.

Thus, administrative and routine tasks, such as scheduling, allocation of resources, and reporting, will within intelligent machines, responsibilities that have long been reserved for humans. For instance, a typical store manager or a lead nurse at a nursing home must constantly juggle shift schedules, accounting for staff members' absense owing to illness, vaction, time or sudden departures. Many of these tasks will be automated by (AI). Imagine (AI) writing management monthly reports, it is not a distant dream. Leading news providers and Wall street banks are now using (AI) report generators to write news and analytical reports by drawing on quantitative data. The associated press, for example, expanded its quarterly earnings reporting from approximately 300 companies to nearly 3,000 with the help of (AI) powered software robots, freeing up journalists to conduct more investigative and interpretive reporting. For another example, Jobalime, a job-placement site, uses intelligent voile analysis algorithms to evaluate job applicants. The algorithm assesses paralinguistic elements of speech, such as tone and inflection, products which emotions a specific voice will elicit, and identifies the type of work at which an applicant will likely excel. In the future , (AI) machines can be applied to assist some kind of office administrative jobs duties. It's attractive to office managers to achieve more accurate judgment to do any administrative matters when who can be assisted from (AI) machines. Thus, managers need to spend time to learn how to apply (AI) machine to assist them to do more accurate judgement, and better informed choices. (AI) robots can be applied to improve the speed quality and cost of available products and services, instead of applying on productivity improvement and administrative improvement aspects. Thus, they may also displace large numbers of workers. This, possibility challenges the traditional benefits model of trying health care and retirement savings to jobs.

In an economy that employs dramatically fewer workers to deliver benefits to displaced workers. For example, the worldwide number of industrial robots has increased rapidly over the past few years. The fall prices of robots, which can operate all day without interruption, make them cost- competitive with human workers. In special consideration, in the service sector, computer algorithums can execute stock trades in a fraction of a second, much faster than any human. As those technologies become cheaper, more capable, and more widespread, they will find even more applicants in an economy.

Consequently, (AI) technology brings unemployed number increasing many businesses continued automating their operations rather than hiring additional workers. A trend among technology companies that receive massive valuations with relatively few workers. For example, in 2014 year Google was valued at $370 billion with only 55,000 employees, a tenth the size of AT & T's workforce in the 1960 year. Hence, if automation technologies like robots and artificial intelligence make jobs less secure in the future, there needs to be a way to deliver benefits outside of employment " flexi security" or flexible security is one idea for providing healthcare, education and housing assistance whether or not someone is formally employed.

In conclusion, (AI) and robots technology will raise unemployment to some occupations when (AI) replaces same industries‘ workers job duties in our societies in the future, but it also create new jobs to raise employment in any related (AI) robots and automated machine products in (AI) manufacturing. (AI) design, (AI) sale self-related industry, when (AI) replaces same industries' workers‘ job duties.

(AI) journalism, media publishing, digital communication technology trend

How to apply (AI) technology in digital communication journalism media, publishing industry? Some scientists indicate future (AI) and digital technology may consist such as: voice driven assistants, emerge. For example, Amazon e book publish applying digital technology and (AI) auto printing technology to sell e books to let readers to listen any e book content by (AI) voice driven speaker when they turn on computer to read e book contents; capable phones start to unlock the possibilities of 3D image of mobile story telling. New smart wearables include ear buds that handle instant translation and glasses that talk and hear. China and India will become a key focus for digital growth with innovations around payment online identity, and artificial intelligence. Thus, future (AI) technology can be applied to 3D image mobile story telling, online payment method to dealt online transaction publishing industry.

Thus, future (AI) technology can be applied to online e book publishing industry to make sound books to let readers feel more attractive . Such as Amazon publish has published sound e books to attract readers to choose to read any its books from online. Also, (AI) technology can also be applied to communication industry. For example, some online pure-play news, opinion and entertainment websites. It is a digital communication media, e.g. online journalism blog (AI) technology can be applied to visual storytellers to let online book readers to enjoy to listen to watch and send any online electronic book contents more attractive. Thus, future (AI) technology will be popular to assist any electronic book publishers to publish visual and sound talking storybook to let readers who can watch motive image and listen and read words from e books more attractive.

Thus, (AI) technology can be applied to internet ecommerce publishing or media industry to help any electronic book publishers to publish sound, image motion electronic book to attract global readers to read, even (AI) technology can be applied to digital entertainment industry, e.g. electronic 3D image virtual video games, computer games. It can be also applied to education industry, e.g. the first true digital native generation and are the native speakers of the digital language of computers to let student to learn different languages or translate words to compare to classroom learning more easily. It can be also applied to communication industry, e.g. (AI) mobile phone. Hence, it seems (AI) technology can be applied to publishing, communication, education , entertainment etc. different industries in the future. (AI) technology will be one kind of tool to satisfy human's daily life needs in the future and these industries has one characteristics is that they need to apply internet to operate to operate to do online business.

Thus, it has three trends of (AI) technology and internet technology need to be linked to achieve one kind of attractive technological business to satisfy client's needs. These three trends as below: All consumer trends involve the internet. It will be many consumer's online habits, shopping, working, socializing, watching TV, studying, travelling, listening. Thus, (AI) music, eating and exercising are just a few examples. This is happening because human usually use mobile broadband or Wi-Fi, rather than cables. Thus, (AI) technology will be applied to mobile to satisfy client's need absolutely.

The mobile phone can be more popular to be used more than computer or laptop tools. The reasons are because women dive the smartphone market by defining mass-market use. But as the speed of technology adoption increases mass market use becomes much quicker then before. Successful new technological products and services , such a (AI) mobile phone products now reach the mass market in popular use. It means that the time period when early adopters influence others is shorter than before. Also, since new products and services increasingly use the internet mass markets are not only faster , but are also more important than ever to consumer themselves. Most internet services become more valuable to individuals when many use them. Thus, it causes why (AI) mobile phone will be popular to be used.

Since, new products and services increasingly use the internet, in the future several trends focus on (AI) smart phone users. Consumers' familiarity with using smartphone apps. Essentially, the technologies will bring other related (AI) and internet service needs, e.g. sound and image emotion e book needs, (AI) mobile communication needs, e-virtual games or e-3D image virtual games etc. entertainment activities needs with such a large part of the world's population now online, it is clear that there is strength in numbers.

Thus, (AI) imagines , if future any (AI) and internet related services or products new technology is easy to use and inexpensive, when the latest products reach the mass market almost as quickly as they reach the early adopters and industry experts. I believe that any (AI) and internet related products or services must be popular to accept to consume for entertainment or useful aim. For example, with major players including Apply, Facebook and Google had invested (AI) technology to develop their businesses. (AI) technology has the potential to disrupt everything in the coming years, from the lives of connected consumers to every industry (AI) will be an alternative route for brands to reach consumers with convincing and relevant messages. Digital technology will assist of the future, then it can improve technology to bring this effect, such as sophisticated software machine learning and speech recognition effective. Hence, Google, Facebook , Yahoo web site service companies can apply (AI) technology to help other companies to advertise their businesses, such as travel, retail, and education etc. industries more attractive. (AI) technology can be applied to internet company to be aware and familiar enough to drive among mainstream consumers, it can create online experience to travel, retail , education and other entertainment needs to online consumers to seek their entertainment needs more easily. Hence, in the future (AI) technology and internet related entertainment service needs will be raised in this (AI) and online consumption market.

- (AI) healthcare service industry development

In the future, (AI) medical internet technology tool can be applied to assist individual's health at the center of their focus, e.g. smartwatch compatible mobile app. patients can let personalized reminders for taking their medication snap pictures of their prescriptions to expedite refills, and scan their insurance card. So that, store clerks are prepared with up-to-date patients' information . (AI) owned health operated technological clinics can help patients to receive treatment for minor illnesses, flu shots, cholesterol screenings and more than a dozen other medical services, all of which can be patients who can't make it to a physical location. (AI) healthcare services organizations can provide various telemedicine services. So, patients can receive care via phone or video chat.

For example, one London-based intelligent Brewing company has developed an (AI) system to continuously collect and incorporate customer feedback, which the system itself uses to brew ne various of the company's beers. Thus, the beer clients can give feedback to talk to the algorithm (AI) machine, whenever or anywhere who're drinking the beer. It is such any healthcare services organizations can apply (AI) machine to collect patient's feedback to talk to the algorithum (AI) machine whenever or anywhere who're eating any medicines. So doctors can know every patient's health conditions any time. If the patients feel uncomfortable, the doctor can know from (AI) machine notification to decide whether the patient needs to eat another new medicine or keep to eat same medicine is better. Hence, (AI) medial internet technological body check report machine will be proper to be needed to serve any hospitals' patients in the future.

However , it brings this question. How can (AI) medical internet technological body check report machine apply to hospital more efficient? The essential new medicine co-workers for the health service digital age health service leaders need apply (AI) medical report machines and artificial intelligence to the newest recruits to the workforce bringing new skills to help health service staffs do new jobs and reinventing what's possible, building the health service workforce for today's digital health service demands for patients. Thus, technology-driven health service model innovation from the health service organization outside in and providing digital health service ecosystems for patients to use the (AI) health service equipment will be popular to be accepted to be used.

● I Robot and internet things future machine men invention

Nowadays, there are some company, which apply internet and (AI) I Robot technology to do any similar human job nature. For fishing industry example, one company, known for creating the Roomba, I Robot is now working with marine conservationists to launch an ocean-patrolling intelligent robot to hunt and manage invasive species, protecting native populations. And evolved industries like precision agriculture are ramping of our increasing population. Area of practice that once seemed impossible to digitize are fundamentally changing because of the impacts of (AI), internet of things capabilities and big data analytics, which have many potentially positive impactions for society.

For textile industry example, automation is nothing new, it has shaped the workplace to replace human jobs to boost productivity in the textile industry. Textile machines have had a generally positive impact over gears, creating value and allowing textile workers to take up more rewarding age will likely continue to create opportunities and lead to new textile industries, companies and textile occupations. It may also compensate for a demographically driven slowdown in the growth of the textile workforce. The future impact of textile (AI) and automatic and internet link is somewhat uncertain. It seems textile industry will be trend to accept (AI) textile workers and internet of thing to replace traditional manual textile workers to produce any shirts, cloths etc. wearing products in factories popularly, during the (AI) textile machine and internet thing technology can be invented to reach the mature stage in the future.

For factory worker transportation job example, they have also expanded their influence, migrating from the factory floor to the service sector and taking the place of humans in a range of activities from financial transactions to transport route optimization. Further (AI) machines and robots are increasingly programmed to learn, meaning they improve with time and undertake cognitive activities. Hence, (AI) machines and internet technology enable automation of work activities to raise factory workers‘ efficient and performances, also factories can reduce manual worker numbers, due to (AI) machine workers’ assistance.

Future, (AI) robotics technologies and internet technique have these different kinds of characteristics: For soft robotics example, it is non-rigid robots construct with soft and deformable materials that can manipulate items of varying size, shape and weight with a single device. For swarm robotics, it coordinated multi-robot systems often involving large numbers of mostly physical robots. For touch/factile robotic example, it robotic body pails (often biologically inspired hands) with capability to sense, touch , dexterity robots example, serpentine robots with many internal degrees of freedom to threat through tightly packed spaces for humanoid robots example, robots physical is similar to human being often bi-pedal that investigate variety capable of performing human tasks , including movement across terrains, object recognition, speech sensing etc. For autonomous cars and trucks example, it is capable of operating with a human pilot, e.g. the unarmed general atomics Predator XPUAV with roughly half the wingspan of a Boeing 737 can fly autonomously for up to 35 hours from take-off to landing, for unmanned aerial vehicles example, flying vehicles capable of operating without a human pilot, the unarmed general atomics predator -XPUAV , with roughly half the wingspan of a Boing 737, and fly autonomously for up to 35 hours from take off to landing, for (AI) chat bots example, (AI) systems designed to simulate conversation with human users, particularly those integrated into massaging apps.

In Dec. 2015. the general service administration of the US Govt. described how it used a chat bot named Mrs. Landingham (a character from the television show the west wing) to help onboard new employees. Finally, for robotic process automation example, class of software robots that replicates the actions of a human being interacting with the user interfaces of both software systems. Enables the automation of many back-office work flows without requiring expensive IT integration . Hence, future (AI) robot machine men will have different functions to be applied to different industries to use in possible.

Statistics Denmark shows that (AI) automation potential robots will influence few jobs are completely automatable , but close to half consists of 40% automatable tasks: It showed example occupations include share of automated, such as brewing machine operators are more than 80%, logging equipment operators are more than 50%, roofers , stock tasks clerks, travel agent are more than 50%, farmers , nursing assistants are more than 30%, physicians, teachers , managers are more than 10%.

For example, humans perform a wide variety of tasks from planting corn to examine spreadsheets, meeting clients and lifting crates in a store. Each of these actions requires a combination of innate or acquired capabilities, internet technique assistance, ranging from social perceptiveness to fine motor skills and natural language understanding. To understand and map automation feasibility by existing technology. Mc Kinsey has developed a framework of 18 technical capabilities that can substitute tasks performed by humans. The capabilities are grouped in five categories: sensory, cognitive, language, social and emotional and physical. So, it seems (AI) robot machine men and internet technique will have possible combination to invent to own human's emotion , language, learning, task skill abilities.

Mckinsey global institute analysis also showed current technologies have achieved different levels of human performance across 18 capabilities include: sensory perception, autonomously infer and integrate complex input using sensors, cognitive capabilities reorganizing known patterns/categories supervised learnings, generating novel, logical reasoning/problem solving, optimization and planning, creative, information retrieval, coordination with multiple agents, output articulation/ presentation, national language processing, social and emotional capabilities-natural language understanding, social and emotion sense, reasoning output, physical capabilities-fine motor skills, navigation mobility. Hence, it seems (AI) robots and internet technological will combine to invent to own human' some skills to replace human to do some kind of tasks in possible.

In conclusion, future (AI) robot and internet will be needed to link to cooperate together to raise human's work efficiency in popular.

How artificial intelligence replaces human job possibility

What is the risk of automation for jobs to replace human job? In recent years, there has been a revival of concerns that automation and digitalization night after all result in jobless future. As I argue, this might lead to an overestimation of job (AI) automate , as occupations labelled as high-risk occupations often still contain a substantial share of tasks that are hard to automate.

For example, when the share of (AI) automatable jobs is 6% in Korea, the corresponding share is 12% in Australia. Differences between countries may reflect general differences in workplace organization, differences in previous investments into (AI) automation technologies as well as differences in the education of workers across countries. I also discover that (AI) automation and digitalization are unlikely to destroy large numbers of jobs. But, however, low qualified labors are likely to raise costs as the (AI) automate of their jobs is higher compared to highly qualified workers.

In fact, (AI) technology will influence some new technology to replace some human's job, such as driverless car, the largely autonomous smart factory , service robots or 3D printing. These technologies are driven by advances in computing power, robotics and artificial intelligence and ultimately redefine what type of human capabilities machines are able to do.

Hence, question brings whether (AI) invention will influence general human jobs to be replaced by (AI) autonomous jobs? Whether will the potential foe automation with actual employment loss? In particular, the technical possibility to use (AI) machines rather tasks need not mean that the substitution of humans by machines actually takes place.

Whether (AI) technology replaces human's some job, it is beneficial to our society or not. Instead, machines are increasingly capable of performing non-routine cognitive tasks, such as driving or legal writing . In particular, advances in the field of machine learning (ML), e.g. computational statistics and visions, data mining, artificial intelligences allow for automating cognitive task, when the use of (ML) in mobile robotics (MR) also allows for automating certain manual tasks. So, it seems, (AI) technology can replace some labor job, e.g. warehouse transportation, even mind's job, e.g. legal writing, driving in possible.

For example, if (AI) automatic non-manual driving can reduce hurt or death risk, it is beneficial to our society, or (AI) automatic robots can more any heavy things (products) in warehouse safely. Then, it can reduce the warehouse labor's bodies hour risk, it is beneficial to the workers. Even, if (AI) robot can write any legal documents, no any word errors in short time. It is beneficial to the law companies , but it also bring unemployment chance, due to these jobs can be replaced by (AI) robots to do in the future. Hence, it will cause some occupation to be disappeared, due to (AI) robots can do our these kinds of jobs in the future.

Frey & Osborne (2013) reported these kinds of occupations will be replaced by (AI) robots in possible. They include computer, engineering, financial, management, legal , art and medium, community service, education, healthcare practitioners and technical service, sales and related, office and administrative support, farming, fishing and forestry, construction and extraction, installation, maintenance, and repair , production, transportation and material moving. It seems our future some professional occupations will have possible to the replaced by (AI) robots to replace, instead of labor jobs. Hence, (AI) robots technology will have much trend to replace high knowledge or low knowledge skillful labors in the future.

In conclusion, it implies that only using information on task-usage at the individual level leads to significantly lower estimates of jobs " at risk", some workers in occupations with according to high automate nevertheless often perform tasks with are hard to automate. Why can (AI) replace human to do some kinds of jobs? (AI) artificial intelligence refers to the ability of a computer or a computer enable robotic system to process information and produce outcomes in a manner similar to the thought process of humans in learning, decision making and solving problem. By extension, the goal of (AI) systems is to develop systems to capable of tasking complex problems in ways similar to human's logic and reasons who feel in our future. Hence, it means future (AI) robots has effort to replace human to do any jobs in possible.

II

(AI) development second stage

(AI) directions for future non-manual
control road vehicles market

Future road vehicle products and technologies must meet social, economic and environmental protection and driving safety goals , and satisfying market requirements for mobility, accident reducing, performance, cost desirability. Thus, (AI) auto-non manual control vehicles need to be followed this direction to invent. To satisfy future driver's safety of needs, enhanced vehicle speed desired functional performance of road transportation system, required and desired technological response, including research needs. It is long term up to 20 years vision, for (AI) auto non-manual research. Thus, (AI) auto non manual vehicle manufacturers need often to revise their (AI) vehicles functions to raise to improve their system performance and driving industry driver's needs, e.g. private drivers need or public transportation driver's need or business client's need. Hence, future (AI) transportation will need have individual driving consumer and business driving consumer both targets.

Thus, future (AI) automation manual control vehicles need to deliver high impact technology solutions to meet social , economic and environmental and safe goals. Engine needs to be improved efficiency , performance, drivability, reliability, durability and speed-to-market together with reduced emissions and cost; hybrid, electric and alternatively fuel (AI) non manual control vehicle technology development, leading to new fuel and power systems, such as hydrogen, fuel cells and batteries, which satisfy future social, economic and environmental and safe goals. Software, sensors, electronics and telematics technology development are needed to be lead to improve vehicle performance, control and adaptability, intelligent , mobility and security, structure and materials technology development, leading to improved safety, performance and leading to flexibility with reduced cost and environmental pollution to achieve the (AI) non manual drivers to feel (AI) vehicle performance, auto control and adaptability is better to compare traditional manual driving vehicles.

In fact, in traditional manual driving market, Japan and USA had had over 80% of world car production by six major global groups. In the future, it is possible only USA can dominate (AI) non manual auto driving vehicle manufacturing market if Japan had no effort to manufacture any (AI) auto non manual control vehicles. So, it means that it is only Japan is USA potential (AI) auto non manual control vehicle manufacturing competitors. Also, it means that it is only USA has effort to export (AI) auto non manual control vehicles to global (AI) auto non manual control vehicle market.

Thus, in long term, (AI) non manual control vehicle product market development, USA (AI) vehicle manufacturers will have these requirement to win new technological competition to traditional manual control vehicle. The requirements include: low cost fuel, low carbon, fuel cell and telematics technologies, the technological roadmap function, such as detailed consideration clear provision of other important areas to the drivers. When the (AI) non manual auto drivers are sitting in the non manual control auto vehicle. Although, who does not need to drive, but who need to know how to go to anywhere by electric road map show clearly. So, the driver won't lose direction and he/she can know the (AI) non manual control vehicles is driving to anywhere in any time, even when who is sleeping.

In the future, the (AI) non manual control vehicles need to be invented to satisfy any business , transportation clients' needs, instead of individual clients needs, e.g. cans, trucks, buses, emergency and utility vehicles, trains, trams etc. Hence, technological road mapping is one important tool to help any business, transportation (AI) non manual control vehicle clients. Technological roadmap is a technique that is used in industry to support strategic planning for (AI) non manual driving vehicles in the future. Electronic road maps generally take the form of multi-layered time based charts, linking technology developments to future (AI) non manual control vehicle market requirements.

Technology road mapping is a flexible technique and the roadmap architecture and process for developing the roadmap most generally be customized to meeting the particular aims. Why technology roadmap will be popular to (AI_ non manual control vehicles. It's advantages include: It is a technology solutions and options that can enable the performance targets to be achieved engine hybrid, electric and alternatively fueled vehicles, software, sensors electronics and telematics, structures and materials design and manufacturing process. It is road transport system performance measures and targets tool, in response to the trends and get (AI) non manual control vehicle drivers to get society, economy, environment protection, low cost driving benefits, also it can help any transportation clients to know how to go to anywhere clearly. Hence, technological road map will be one good tool to assist (AI) non manual control auto vehicle to develop future road driving market.

Reference(source)

Frey & Osborne (2013), The future of employment: How susceptible are jobs to computerization? University of Oxford.

Mckinsey Global Institute Analysis

Statistics Denmark, Global automation impact model, Makinsey analysis

(AI) -driven automation
industry development

(AI) -driven automation industry will create wealth and expand economy growth to any countries, but it will be accompanied by changed in the skills that workers need to learn. One of main ways that technology increases productivity is by decreasing the number of labor hours needed to create a unit of output. It implies (AI) technology will influence low educated and low skillful labor number to be decreased (reduction employment number).
In contrast, technological change tended to work in a different direction throughout the nowadays. The advance of computer and the internet raised the relative productivity of higher skilled workers. So, routine-intensive occupations that focused on predictable tasks disappearance, such as switch board, operators, filming checkers, travel agents and assembling line workers etc. were particularly replaced by new technologies.
However, today, it may be challenging to predict exactly which jobs will be most immediately affected by (AI) driven-automation. The reason is because (AI) is not a single technology, but rather a collection of technologies that are felt unevenly through the economy to influence job changing both negatively and positively. In positively view point, (AI) driven-automation will make many workers more productive and increase demand for certain skills. Consequently, new jobs are likely to be directly create in areas , such as the development and supervision of (AI) as well as indirectly created in a range of areas throughout the economy as higher incomes lead to expanded demand. Otherwise, in negatively view point, many traditional human needed (demand) skillful jobs will be threatened by automation are highly concentrated among lower-paid, lower-skilled and less -educated workers. It means automation will cause pressure on demand for this group, pressure and

employment, if (AI) can replace the low skilled and less educated workers' jobs. Thus, (AI) will have negative influence to impact on the labor market.

(AI) capabilities will enable automation of some tasks that have long required human labor. Why can (AI) replace some simple human jobs? For example, advances in robotics are expanding machines' abilities to interact with and sharp the physical world. Combined , (AI) and robotics will give rise to smarter machines that can perform more sophisticated functions than ever before and brings more advantages that humans have exercised. This will permit automation of many tasks now performed by human workers and could change the shape of the labor market and human activity.

5.1 How (AI) influences labor market

Today, it may be challenging to predict exactly which jobs will be most immediately affected by (AI)-driven automation. Because (AI) is not a single technology, but rather a collection of technologies that are applied to specific tasks.

Some specific predictions are possible based on the current (AI) technology. For example, driving jobs and house cleaning jobs, bank counter service jobs, telephone enquiry service operators. Restaurant cooking jobs, simple accounting record service jobs etc. that require relatively less education to perform. Advancements in computer vision and related technologies have made the feasibility of fully appear more likely, potentially displacing some workers in driving-dominant professions. Seemingly similar robot, for which the operational tasks is less specific of navigating to a specific destination when following a set of given rules and preserving safety.

In the future, the effects of (AI) on the labor market in the decade ahead will continue the trend toward skill-biased change that computerization and communication innovations have driven in recent decades. Thus, some human driving occupation will be disappeared or replaced by (AI) automation driven. For example, bus drivers, light truck or delivery services drivers, heavy and

tractor-trailer truck drivers, school drivers, tax drivers, travel bus drivers.
However, (AI) technology could enable some workers to focus time on other job responsibilities, boosting their productivity, and actually raised wage growth among those still holding the reshaped jobs. For example, salespeople, who currently spend a considerable amount of time driving could find themselves able to do other work when a car drives them from place to place, or inspectors and appraisers could fill out paperwork, when their car drives itself. This (AI) -driven technology should make these workers more productive, with (AI) -driven technology serving as a complement, not a substitute. New jobs will also likely be created, both in existing occupations cheaper transportation costs with lower prices and increase demand for products and all the related occupations, such as service and fulfillment, and in new occupations not currently foreseeable.
What kind of jobs will be created by (AI) technology? Predicting future job growth is extremely difficult, due to it depends on technologies or substitute for existing today as well as they may complement or substitute for existing human skills and jobs. However, (AI) will also lead to substantial indirect job creation to the degree it raises productivity and wages, it may also lead to higher consumption that would support additional jobs from high-end draft production to restaurant and retail. The future(AI) " augmented intelligence", the technology's role is as assisting and expanding the productivity of individuals rather than replacing human work. Thus, based on the biased-technical change framework, demand for labor will likely increase the most in the areas where humans complement (AI) automation technologies. For example, (AI) technology , such as IBM's Watson may improve early detection of some cancers or other illnesses, but a human healthcare professional is needed to work with patients to understand and translate patients' symptoms, inform patients of treatment options, and guide patients through treatment plans. Shipping companies may also partner workers who pick up and

deliver products over the last feet with (AI) enabled autonomous vehicles that move workers efficiently from site to site. In such cases, (AI) augments what a human is able to do and allows individuals to either be move effective in their specially task or to operate on a larger scale. Thus, it seems (AI) technology will also create new jobs, raise productivities and workers' efficiencies.

5.2 Redefining management in
the workforce of artificial intelligence

In the future, due to artificial intelligence influences to some kind of human jobs nature. So, the kind of human jobs of management methods will also need to change to adapt the artificial intelligence technology input to their organizations. It will cause challenges for every executive and manager if who won't have effort to manage their teams how to apply artificial intelligence technology to work efficiently and easily. For example, division of labor will change among humans and machines will increase. Thus, companies will have to adapt their training performance and talent strategies how to emphasize on work that how to make human judgment and skills and experimentation. Thus, (IA)'s greatest impact will be on administrative coordination and control tasks, such as scheduling , resource allocation.

In fact, mangers will encounter this challenges: How to apply human experience and expertise to judge critical business decisions and practices when the information available is insufficient to suggest a successful course of action? Due to this kind of work will require new skills and mindsets. I shall indicate these change management methods to adapt (AI) technology. Such as: administration and routine tasks, scheduling , allocation of resources and reporting will fall within the intelligence machines, responsibilities that have long been reserved for humans. For example, a typical store manager or a lead nurse at a nursing home most constantly arrange shift schedules, accounting for staff members' absences owing to illness, vacation time or sudden departures.

Thus, the managers need to learn how to arrange new division of labor within the organizations after (AI) technology had been implemented to the organization. Artificial intelligence is currently influencing into once considered exclusive to humans: assessing and acting on human emotions and personality traits. The influences to managers need to change their strategies to adapt (AI) technology implements include such as below:

Firstly, managers need to spend the bulk of their time on coordination and control tasks from intelligent system implements. Their time spending on these major three aspects from impact of intelligent system: coordinate and control, solve problems and collaborate and people and community , strategy and innovation three aspects. Thus (AI) will influence managers need to change their judgment method to teach whose teams how to adapt the (AI) system operations in any organizations.

Secondly, (AI) will influence top, middle and low level management needs to change to adapt the (AI) technology operations to any owned (AI) technology organizations in the future. Intelligent machines must be trained in context. Just like humans , on-the-job training is a requirement for such machines because they typically arrive with only very general capabilities. To get the most from (AI), managers at all levels must participate in the instructional experience and in the learning process and provides managers' familiarity with such systems on these aspects, e.g. How the system works and generate advice, how the system has a proven track record , how the system provides convincing explanations , how the system can make simple rule- based decisions.

Thirdly, managers need to learn how to make judgment more accurate (AI) systems assistance. Although (AI) will invariably take on more routine work and even augment human decision-making, it won't judgment work, the application of human experience and expertise to critical business decisions when the information available is insufficient to suggest a successful course of action or reliable enough to suggest an obvious course of action. For a sense of the nature of judgment work, consider big data marketing and sales analytics. Such analytics often provide insights that can inform promotional campaigns, including predicting which promotions will generate desired sales brand further into the future, marketing executives need use judgment, combining analytics with their own and others' insight and experience.

The application of experience and expertise to critical business decisions and practice represents the real value of human judgment. But, when artificial intelligent machines are implemented to any organizations to assist the low, middle and top level management to make any business judgment. These forms of judgment work that managers can gather data interpretation, idea development more absolute from (AI) machine assistance. Thus, why these level management executives need to learn how to apply (AI) machines to help them to make any business judgment more accurate.

5.3 How (AI) influences organizational change

Consequently creative and social intelligence will be in even greater demand as (AI) makes in management and the workforce. This development will represent a long term trend in labor markets , one characterized by intensifying demand and reward for social skills with a growing desire for creative capabilities, managers will seek to fashion of ideas and hypotheses from inside and outside of the enterprise to shape solutions to their most pressing business problems. Thus, (AI) will influence overall organizational team members who have chance to participate any decision to make more accurate business judgment.

Many managers mistakenly view judgment work as only an individual discipline, failing to appreciate that it can also involve decide interpersonal and organizational practices. In more complex settings, judgment is typically a collective outcome of individuals' and teams' diverse perspectives, insights and experiences. And often , the resulting choices are better informed than decisions that an individual would have arrived at on his or her own.

Thus, when any organizations apply (AI) technology to assist managers to gather data and ideas to make any judgment. In these cases, organizations can create the conditions for effective collective judgment by establishing structures , such as " shadow advisory boards" that prompt managers and employees to source and synthesize multiple perspectives. Thus, a traditional organization (firm) might freshen its thinking is t put together a shadow advisory board, comprised of young, digital people who can apply (AI) machine assistance to make judgment work more accurate whether related to people development, problem-solving or strategizing and innovating for considerable degrees of creative and social intelligence.

Thus, on the one hand, (AI) technology machine augmentation and automation can give these advantages to human (organization managers) , e.g. developing people and community, solving problems and collaborating, coordinating and controlling work, shaping strategy and leading innovation. Besides, on the other hand, the next generation managers need have these individual attitude to treat intelligent machines to be as colleagues.

When, judgment is a human skill, intelligent machines can accelerate human learning that supports it, assisting in data -driven simulations, scenarios and search and discovery activities. Focuses on judgment work, some decisions require insight beyond what data can tell them. This is the sweet sport for human judgment, the application of experience and expertise to critical business decisions and practices. Thus, managers will also need to find ways to learn how to use digital (AI) technologies to tap into the knowledge and judgment of partners, customer external stakeholders and role models in other industries after the (AI) machine had been implemented to the organization.

5.4 Future works change:
Automation, employment
and productivity

Human future " micro to macro" industry trends will be affected business strategy and public policy by (AI) technology. In the future (AI) technology will influence those six themes: productivity and growth, natural resources, labor markets, the evolution of global financial markets, the economic impact of technology and innovation and urbanization. However, (AI) technology will bring economic benefits of tackling gender inequality, a new global competition, Chinese innovation and digital globalization.

Nowadays, advances in robotics artificial intelligence, and machine learning are in a new age of automation, as machines match or outperform human performance in a development to any countries. For example, automation of activities can enable businesses to improve performance by reducing errors and improving quality and speed, and in some cases achieving outcomes that go beyond human capabilities. For example, some research indicated automation could raise productivity growth globally by 0.8 to 1.4 % annually; more than 2,000 work activities across 800 occupations. When less than 5% of all occupations can be automated using demonstrated technologies about 60% of all occupations have at least 30% of constituent activities that could be automated. Many occupations will change that will be automated away: Activities most susceptible to automation involve physical activities, in highly structured and predictable environments, as well as the collection and processing of data. They are most prevalent in manufacturing , accommodation and food service and retail trade and include some middle-skill jobs. For example, such as natural language processing is a key factor. Beyond technical feasibility, the cost of technology competition with labor including skills and supply and demand dynamics, performance benefits including and beyond labor cost savings, and social and regulatory acceptance will be affected by (AI) automation technology. Thus, (AI) automation will impact to influence global employment in those aspects as below:

Firstly, assuming that people are displaced by automation will find other employment. The anticipated shift in the activities in the labor force is of a similar order as the long-term shift away from agriculture and decreases in manufacturing share of employment. Both of manufacturing and agriculture industries which would be accompanied by the creation of new types of work not foreseen at the time.

Secondly, for business, the performance benefits of automation are relatively clear. Thus, the businessmen have opportunities for their micro economies to benefits from the productivity growth potential and macro economies to benefit to encourage continued progress and innovation , investment and market incentives. At the same time, employers must innovate policies to help workers and institutions adapt to the impact on employment.

This will likely include rethinking education and training, income support and safety nets , as well as support for those dislocated, when employees need to leave themselves homes to move to other cities to learn new (AI) automation works. Thus, individuals in the workplace will need to engage move comprehensively with machines as part of their everyday activities, and acquire new skills that will be in demand in the new automation age. Consequently , the scale of shifts in the labor force over many decades that automation technologies can be a similar order to the long -term technology -enables shifts in the developed countries' workforces away from agriculture in the 21 th century. Those shifts did not result in long-term mass unemployment because they were accompanied by the creation of new types of work not foreseen at the time. However, human will still be needed in the workforce when the total productivity gains are caused by (AI) technology.

5.5 What occupations will be influenced by (AI) technology.

In the future, scientists predict that these occupations will be influenced by (AI) technology mostly. They include : retail salespeople, food and beverage service workers, language or translation teachers, health practitioners. Since these work activities have a more relevant occupations are made up of a range of activities with different potential for (AI) automation . For example, a retail salesperson will spend more time interacting with customers, stocking shelves , or ringing up sales. Each of these activities is distinct and requires different capabilities to perform successfully.

Thus, these job activities have similar simple control characteristics.

Simple activities include greet customers, answer questions about products and services, clean and maintain work areas, demonstrate product feature process sales and transactions. All these activities can have similar simple activities in order to (AI) machines can be learn how to do these activities from (AI) technology . For example, the capability perception includes sensory perception, cognitive capabilities, such as retrieving automation, recognizing known patterns(supervised learning), logical reasoning problem solving.
Thus, (AI) machine is such human, which has feeling and emotion, such as social and emotional sensing, judgement reasoning methods, natural language understanding and physical capabilities, such as mobility , navigation, gross motor skill, fine motor skills. It seems that the future, (AI) human invents machines which will have these human characteristics to do human similar behavioral job duties more easily and efficiently. It implies these above human occupations will be replaced by (AI) human invention machines in the future. Due to (AI) creation, it is possible to cause unemployment number of these above workers will increase because (AI) machines can do their similar job behavioral activities. Consequently, employers won't need to employ many of these skillful labor. Otherwise, they can buy less number (AI) machines to attempt to do whose job activities more easily and efficiently. So, it seems (AI) machines will have more high work performance to replace these occupation workers' work performance. Finally, these occupation worker unemployment number will only increase when the (AI) machines had been invented to achieve to do their work behavioral activities absolutely success in the future.

5.6 Whether (A) technology machine labor will replace human worker more or assist human worker more

There is no single agreed definition of a robot how outcome of a task that is completed without human intervention. When some definitions require the task to be completed by a physical machine moves and respond to its environment, other definitions use the

term robot in connection with tasks completed by software , without physical embodiment.
However, to answer the question : Whether (AI) technology machine labor will replace human worker more or assist human worker more. I shall indicate some examples to let readers to judge whether (AI) technology can create new jobs or reduce old jobs.
Firstly, I shall explain what (AI) function is. (AI) is a service robot that performs useful tasks for humans or equipment excluding industrial automation application . Thus, the classification of a robot into industrial robot or service robot is done according to its intended application. It is also a personal service robot or a service robot for personal used for a non commercial task, usually by lay persons . Examples are domestic servant robot, and pet exercising robot. It is also a professional service robot or a service robot for professional used for a commercial task, usually operated by a properly trained operator. Examples, are cleaning robot for public places, delivery robot in offices or hospitals, fire-fighting robot, rehabilitation robot and surgery robot in hospitals. Thus, these functions will be future (AI) application to our daily life necessaries or business necessaries.
However, some authors agree (AI) will bring negative outcomes of automation, due to raise competiveness, reduce human job nature. Otherwise, other authors argue (AI) will bring positive outcomes of automation, due to raise productivities, job creation, assist humans work.
On the positive outcome hand, robots can increase productivity . This is particularly important for small-to medium sized businesses both are in developed and developing countries economies. It also enables large companies to increase their competitiveness through faster product development and delivery. Increased use of robot is also enabling companies in high cost countries to re shore, or bring back to their domestic base parts of the supply chain that will have previously outsourced to sources of cheaper labor. Currently , the greater threat to employment is not a automation, but an inability to remain competitive. Automation has led overall to an increase in

labor demand and positive impact on wages. The reason is that the middle-income/middle-skilled jobs have reduced as a proportion of overall contribution to employment and earnings leading to fears of increasing income inequality, the skills range within the middle income bracket is large. Thus, robots are driving an increase in demand for workers at the higher -skilled and with a positive impact on wages. This issue is how to enable middle-income earners in the lower-income range to unskilled or retain. Finally, the (AI) positive impact supporter who argue the future will be robots and humans can work together.

However, on the negative outcome hand, robots can substitute labor activities, but don't replace jobs. They believe that less than 10% of jobs are fully automatable. Increasingly , robots are used to complement and augment labor activities, the net impact on jobs and the quality of work is positive. Automation can provide the opportunity for humans to focus on higher-skilled, higher-quality and higher-paid tasks. Robots can improve productivity when they are applied to tasks that which perform more efficiently and to a higher and more consistent level of quality than humans. For example, increased productivity is enabling some firms, such as Whirlpool, Caterpillar and Ford Motors company in the US restructure their supply chains, bringing back parts of the manufacturing process to the country of origin. Thus, productivity gains due to robotics and automation are important not just at the company level, but also for build industry and nation competitiveness.

I suppose that productivity can be raised. What are the impacts of robots on employment? Firstly, the main focus of development has been on personal entertainment, which does not drive worker productivity (manufacturing production). When the internet (information and communication technology (ICT)) innovation. This is borne and by findings that manufacturing productivity, which has been driven by innovations in automation rather than consumer technologies, has government strongly than productivity in the services sectors of the economy in most nature economies.

It seems (AI) automation will create many jobs in internet communication entertainment game industry. For example, many young people like to use internet to play any electronic games from computer or mobile at home or outside home conveniently. Thus, (AI) automation will increase demand to be invented to any new entertainment game from internet channel. It will need to employ many (AI) entertainment game inventors to create many automation entertainment games. Thus, (AI) automation in internet entertainment game industry will need human (AI) entertainment game inventors to invent the knowledge-based capital of (AI) automation entertainment games. The (AI) entertainment game inventors will need own research and development skills, form specific skills, organizational know-how skills, databased knowledge, design and various forms of intellectual property to do these (AI) automation entertainment game invention occupations in the future.
International Federation Of Robotics(2016) indicated that China will be as a major robotics manufacturer and user of robots, benefiting from jobs created by robot manufacturing and productivity gains from robot use. Chins had sold of robots to any one single market every year since 2017 year. The Chinese government has included a focus on robotics in its 10 year strategy. In order to achieve its target of a robot density of 150 units per 10, 000 workers by 2020 year. Thus, Chinese companies will have to install around 650,000 new industrial robots between 2016 to 2020 year, 2.5 times more than installed globally in 2015 year.
Hence, China (AI) manufacturing industry will need to employ many workers . It implies (AI) manufacturing industry will create many new occupations in China. Also, ministry of economy, trade and industry (2015) also showed that Japan currently has the largest stock of industrial robots in operations, primarily in the automation industry. Driven by a rapidly aging population and low productivity rates, the Japanese government has sights on a 20-fold increase in the use of robots in the non-manufacturing sector and a three-fold growth rate of labor productivity in the service sector

both by 2020 year. Thus, it also implies Japan will need many robots to be provide to service industry. Due to robots will provide to serve any businessmen's clients. Thus, it is possible that the service workers won't be dismissed as well as it is depended on the serving job nature to decide whether Japan's service workers can still serve to their employer when the service (AI) robots are applied to whose employers.

Consequently, it seems that (AI) can create employment, Ministry of economy, trade and industry (2015) showed that such as China will develop the major (AI) automation manufacturing industry. The (AI) employers will need to employ many workers to manufacture any these different kinds of (AI) robots to satisfy China or overseas individual or business buyers needs. But, (AI) can also cause unemployment to the low skillful service workers. Such as if Japan some service businesses choose to buy any (AI) service robots to replace their service staffs to serve their clients. It is possible that the service staffs will be dismissed, due to (AI) robots can do such as their same service job duties to achieve better service performance. Thus, today, it is increasingly common for people to use robots in various situations at home and in retail stores, hotels and hospitals these service industries. Robots are classified into server types based on their functionality (service and utility robots or those designed to communicate with humans) and appearance (humanoid robots or mechanical robots). The type of robot, to which each country allocated particular importance in the advance of robotics, reflects the sense of values and preferences of its population. Thus, if the country has high population needs to use robots, then they will influence either more new jobs creation or more old job loss in the country's (AI) manufacturing or (AI) service industries both. For example, Japan respondents often associate the term " robot " with humanoid robots that can communicate with human and they have a high level of familiarity with robot. The US has the highest level of robot utilization at home and in retail stores with its people being the most enthusiastic about the future use of robots. Germany shows a strong tendency to consider robots for industrial purposes and its people feel strong effort to the presence of robots in their households.

In conclusion, to judge whether how (AI) will influence the country's employment to be better or worse. It will depend on the country home buyers (users) or business buyers (users) how to use (AI) for their daily needs. If the country , such as US retail stores need to use (AI) , it will have possible to reduce some or many retail service workers. Even, if the country , such as Japan has many home users need to use (AI) , it will not influence the employment market. Otherwise, it will raise (AI) salespeople numbers. Even, if the country, such as Germany and China will have many (AI) manufacturers, then it will create many (AI) manufacturing occupations for these (AI) manufactory workers.

Consequently, (AI) robots manufacturing and service needs will have positive or negative impact to any country's employment. It will depend on the (AI) service provision and service workers' job nature as well as the manufacturing workers of (AI) knowledge level to decide their employment chance in their country's employment market.

What does artificial intelligence(AI) mean

- What (AI) function is?

Some scientists explain that artificial intelligence means which is an expert system, computer software that embodies a portion of the specialized knowledge of a human portion in a specific, narrow domain, owns decision making ability of human expert. The (AI)technology is based on the premise that what makes a person an expert is years of experience that enables who recognizes certain patterns in a problem as being similar to pattern. For example, in the future artificial intelligence system can be applied to control air traffic, design to computer configuration, medical diagnosis, instruction/training, speech/interpretation, monitoring to (nuclear plant), planning to mission, factory scheduling, prediction weather, repairing telephone, automatic driving etc. different industries.

Artificial intelligence characteristics include: creative, adaptive , common sense, fact processing, quick replication, broad focus permanent and consistent skill. Otherwise, traditional computer expert system disadvantage includes perishable, unpredictable, slow reproduction, expensive, slow reproduction, slow processing lacks inspiration, needs instruction, narrow focus only machine knowledge. So, artificial intelligence is a branch of computer science devoted to creating computer to influence software and hardware to attempt to create human intelligence or human intelligent behavior. It is learning from experience, responds flexibility in situation that are, new or not anticipated.

Thus, (AI) can be learnt programmed knowledge to solve problems, using reasoning in solving problem, understanding and inferring facts and rules, recognizing the relative importance of different elements in a situation. In summary, artificial intelligence is concerned with two basic ideas mainly: The first idea, it involves studying the thought processes of humans to understand what intelligence is; the second idea, it deals with representing thought processes using companies to create artificially intelligent entities for testing the theories of intelligence.

- Can (AI) impact human job nature?

Human need concern this question: Will artificial intelligence (AI) reduce some human jobs in order to instead of replacing machines to do? Due to artificial intelligence is the ability of machines to do thing, that people would require intelligence. For example, artificial intelligence machine man driving(self-driver), it (AI) machine man driving research is an attempt to discover and describe aspects of human intelligence that can be simulated by driving machine functions. Alternatively, (AI) mathematical research may be another viewed as an attempt to develop a mathematical theory function to describe the abilities and actions of things (natural or man-made) exhibiting intelligent behavior and server as a design of intelligent calculation machine function.

Why do humans need artificial intelligence machines to instead of traditional human service job? For example, can artificial intelligence machine man (self-driving) driver drive to replace human driver? I shall compare the differences between humans and computers : The characteristics of humans are good at recognizing various things, either seen before or not, recognizing the relationship patterns between things. Human thinking is common sense reasoning, combining all types of sensory input, acting appropriately in novel situations, learning new things and changing behavior patterns, making decisions , even when given incomplete information, working with noisy, incomplete information gathering behaviors . However, characteristics of computers are good at: The tasks humans do naturally are extremely difficult for a computer program as intelligent, which must be able to do the same kind of tack as humans do naturally.

Hence, (AI) is an combination of many different success and technologies: Linguistics - computational and socio, philosophy-logic, philosophy of mind and of language, electronical engineering -image and speech processing, pattern recognition, robotics, machine learning, neural networks, optimization scheduling, management information system and decision making. So, it is possible that (AI) can impact human job nature to instead of human working behavior in the future.

● How can human society job nature
to be changed to artificial intelligent society?

From the first intelligent perspective reason view point, artificial intelligence is making machines " intelligent" acting as humans expect people to act. Artificial intelligence has ability to distinguish computer responses from human responses, it owns knowledge to solve expert problem. From another research perspective reason view point, artificial intelligence is the study of how to make computers do things which, at the moment, people do better (Rich & Knight, 1991, p.3).

(AI) researchers are native in a variety of domains, e.g. formal tasks (mathematics, games), tasks (perception, robotics, natural language, common sense reasoning), expert tasks (financial analysis, medical diagnostics, engineering, scientific analysis and other areas).

From the second business perspective reason view point, (AI) is a set of many powerful tools, and methodologies for using those tools to solve business problems. From a programming perspective reason view point, (AI) includes the study of symbolic programming problem solving and search .

From the third human technological perspective reason view point, today's computer can do many well-defined tasks, for example, arithmetic operations, are much faster and more accurate than human beings. However, the computers' interaction with their environment is not very sophisticated yet. How can human test whether a computer has reached the general intelligence level of a human being? Can a computer convince a human interrogator that it is a human? But before thinking of such advanced kinds of machines, human will start developing our own extremely simple " intelligent" machines.

So, it is possible that human society job nature will to be changed to artificial intelligent society when (AI) technology is developed to the mature stage in the future.

- Why does human need artificial intelligence machines?

One of major division in (AI) is between humans who think (AI) is the only serious way of finding out how we (human) work and human who want companies to do very smart things, independently of how we (human) work. This is the important distinction between cognitive scientists vs engineers. One of another major division in (AI) is between symbolic (AI), which represents information through symbols and their relationships. Specific Algorithms are used to process these symbols to solve problems or deduce new knowledge and connectionist. So (AI) , which represents information in network. Biological processes underlying learning, task performance and problem solving are imitated from human mind behaviors.

Thus, it is possible that artificial intelligence machines can do the better judgicious behavior to compare human.

- How does artificial intelligence influence future working changing in automation employment and productivity aspects?

In the automation changing influence aspect, as companies

increasingly use robots on production lines or algorithms to optimize their logistics manage inventory, any carry out other core business functions. Technological advances are creating a new automation age in which ever-smarter and more flexible machines will be deployed on an ever larger scale in the marketplace. However, researching artificial intelligence with how influences human working nature. We need to answer these questions: How will automation transform the workplace? What will the implications for employment? And what is likely to be its impact both on productivity in the global economy and on employment?
Advances in robotics, artificial intelligence, and machine learning are growing in a new age of automation as machines match or outperform human performance in a range of work activities, including ones requiring cognitive capabilities. What factors are determined the changing in workplace adoption by artificial intelligence innovation? What advantages are automation? Automation of activities can be enabled businesses to improve performance by reducing errors and improving quality and speed, and achieving outcomes that go beyond human capabilities.
Some scientists indicated based on their scenario modeling. They estimated automation could raise producing growth globally by 0.8 to 1.4 percent annually. Almost, the activities people are paid almost $16 trillion in wages to do in global economy have the potential to be automated by adopting currently demonstrated technology. According to their analysis of more than 2,000 work activities across 800 occupations. When less than 5% of all occupations have of least 30% of activities that could be automated. They also indicated that technical economic and social factors will determine automation. Continued technical progress, for example, in areas such as natural language processing is a key factor beyond technical feasibility , the cost of technology, competition with labor including skills, and supply and demand dynamics, performance benefits including and beyond labor cost savings and social and regulatory acceptance will affect (alter) the scope of automation.
Other some scientists also indicate U.S. country for example, the

anticipate shift in the activities in labor force of a similar order of magnitude as the long term sight away from agriculture and decreases in manufacturing. Share of employment in the United States both which were achieved. So, those factors can influence why artificial intelligence technology needs. So, it is possible that future agriculture and manufacturing both industries will apply (AI) technology manufacturer-kind of job nature to raise productivity instead of farmers, fruit picking workers, farming transportation labours as well as factory manufacturing workers and supervisors etc. human-kind of job nature.

- Is artificial intelligence possible to replace labor ?

Not just intelligence, but also debating, if machines are capable of having a conscious minds. Artificial intelligence has those characteristics as below:

On functionalism aspect, artificial intelligence inputs mental states, sensory inputs, (beliefs, desires being in pain feeling) and behavioral outputs. Since mental states are identified by a functional role, which are thoughts to be manifested in various systems. Even, perhaps computers which are physical devices with electronic substrate that inform computations on inputs to give outputs similar to brains which are artificial intelligence composed of part any intrinsic relationship to each other. Thus, artificial intelligence activities is not the whole itself, but into parts or on external influence on the parts.

On dualism aspect, artificial intelligence is a set of views about the relationship between mind are matter. On materialism aspect, it builds the only thing that exists is matter, including consciousness. On biological naturalism aspect, it is similar a human brain than feels pains makes mental situation. So, artificial intelligence is similar biologist which might to be excited to human labor work.

Hence, it seems artificial intelligence can change (alter) or replace human labor work of nature in possible in the future.

- Can (AI) technology replace human labour nature of work?

On technological innovation reason view point, the history development of artificial intelligence studying the intelligence is one of most ancient scientific discipline. The history development of artificial intelligence what aims to achieve human use to sense, learn remember and think, logic probability, decision making and calculation develop from mathematics, instead of replacement human labor functions.

Artificial intelligence history development aim is the scientific analysis of skills in connection and practice with the appearance of computers from 1950 year beginning. The artificial intelligence (AI) can deal with the ultimate challenges. How can (either biological or electronic) mind sense, understand and manipulate a world that is much simple and more complex than itself? And what if would human like to construct something with such capabilities?

The general-purpose software of the early period of (AI) were only able to solve simple tasks effectively and failed when which should be used in a wider range or an more difficult tasks. One of the sources of difficulty was that early software had very few or mix knowledge about the problems which handled, and activities successes by simply syntactic manipulation. Moreover, the other difficulty was that many problems that were tried to solve by the (AI) were untreatable.

The early (AI) software whether trying step sequences based on the basic facts about the problem that should be solved, experimented with different combinations till which found a solution. From the end the 1960 year, developing the so-called expert systems were emphasized. These systems had (sue-based) knowledge base about the field which handled. Till to the beginning of the 1970 year, (Prolog) the logical programming language was born, which was built in the computation realization of a version of the resolution calculus. (Prolog) is a remarkably prevalent tool in developing expert systems (on medical, judiciary and other scopes), but natural language parsers were implemented in this language. Then, in 1981 s, the Japanese announced the fifth generation computer system project, a 10 years plan to build an intelligent computer system that use the (Prolog) language as a machine code. Nowadays, (AI) can be applied any industries, such as car manufacturing industry can use (AI) technological machine-men manufacture car, instead of replacing human labors in factory. Even, in the future, using (AI) machine-men drivers can drive any private cars or public transportation tools, instead of replacing human drivers, e.g. bus, train, tram, ferry etc. Also in the future, machine-men can replace housewives to serve families to do housekeeping clean job , e.g. cleaning toilets, bathrooms, kitchens, even cooking functions at home. So (AI) machine-man can reduce housewives works at home. Moreover, (AI) machine man can take care old people , when who are living at homes or elder care centers.

So, it seems artificial intelligence (AI) will be possible developed to manufacture a new generation machine-man to assist (serve) families to do any simply cleaning or cooking jobs at homes. Moreover, the overall demand of (AI) general social needs will also rise, such as security, driving transportation tools, restaurant cleaning, elder centers care service etc. So, it seems that individual or families or social needs of (AI) will be increase in the future. Thus, it will influence macro economy growth (GDP) if there are large house family consumer group and hotel or bus or taxis or ferry etc. different business consumer group demand any artificial intelligence machine numbers increasing. Then, the artificial intelligence products and material manufacturers must need to buy many artificaial intelligence materials to produce any kinds of artificial intelligence machines to prepare to satisfy consumer individual needs. Consequently, macro economy will grow to the owned artificial intelligence development countries, e.g. US, China, UK.

- Why can artificial intelligence satisfy human needs?

First, On machine-man satisfactory demand aspect view point, it makes computers that think, it is the automation of activities. We associate with human thinking: like decision making, learning. It is the act of creating machine that perform function that require intelligence when performed by people. It is the study of mental faculties through the use of computational models. It is the study of computations that make it possible to perceive, reason and act.

It is a branch of computer science that is concerned with the automation of intelligent behavior. It is anything in computing service that human don't yet know how to do property.
Second, on thought aspect artificial intelligence means systems thank think like humans, systems that think rationally.
Third, on behavioral aspect, artificial intelligence systems that act like human and that systems act rationally. However, the basic objective of (AI) is to represent human's thought processes in computation . These machines are supposed to exhibit behavior that. It is performed by a human being, would be considered intelligent. However, some authors feel (AI) has disadvantages, such as it is not creative, it is excited in the use of sensory devices, it can't make use of a very wide context of experiences and it does not use common sense.
For speech recognition and understanding function needs example, (AI) can be applied in speech recognition and understanding function, which (AI) speech or voice recognition is a data input method. For example, the computer recognizes and understands one (or a few) word commands. Speech understanding on the other hand is the computer's ability to understanding a spoken language. That is , the computer understands the meaning of sentences, an paragraphs through (AI).
So, (AI) can be attempted to learn human language how to speak. It is similar to translate human language skill, instead of actual human speaking skill. Also, (AI) can assist handicap learning or language student how to listen different languages by machine-man sounds from computers more accurately.
So, it seems that it (AI) can replace human language teachers speaking function and can change teaching language nature of job in language speaking and listening education industry.

- Is artificial intelligence one good choice for human future technological benefit?

Nowadays, new technology development is popular. However, artificial intelligence is one kind of new technology choice among different technologies innovation. So it brings this question: Is artificial intelligence technology value to invest? To answer this question. I shall indicate some other new technology developments to compare (AI) technology development to judge which has urgent needs to achieve human expectation nowadays.

For example, why is green peace interested in new technologies? New technologies features prominently in our ongoing campaigns against genetic modified crops and number power. However, which are also an integral part of our solutions to environmental challenges, including renewable energy technologies, such as solar, wind and wave (water) power energy as well as waste treatment technologies, such as mechanical, biological treatment.

It seems humans need concern how to apply (AI) technology to solve environment pollution challenges in our future. So, environment protective, agriculture, natural energy technology will be popular demand to attempt to apply (AI) technology to solve their challenges or apply (AI) to assist to develop their industry.

What is relationship between
(AI) and economy growth?

- How can artificial intelligence technology influence economy?

Advances in artificial intelligence (AI) technology and related fields have opened up new markets and new opportunities progress in critical areas, such as health, education, energy, economic development, social welfare and the environment pollution.

(AI) automation will continue to create wealth and expand the global economy development in the future. However, when many will benefits that growth won't be costless and will be accompanied by changes in the skills, that workers need to increase productivity in the economy and structural changes in the economy. So, in the skills that workers need to succeed in the economy and structural changes.

I shall indicate why aggressive policy action will be needed to help Americans who are disadvantaged by these changes , due to (AI) technology is caused. For automation industry change example, artificial intelligence (AI) capabilities will enable automation of

some tasks that have long required human labor. These artificial intelligence technology introduction can increase new opportunities for individuals. The economy and society, but (AI) has also the potential to disrupt be current livelihoods of many Americans. However, (AI) leads to unemployment and increase in inequality over the long run depends not only on the (AI) technology itself, but also on the institutions and policies that are changed.
Thus, it is possible that (AI) technology will raise some countries unemployment number if the employer apply (AI) technology workers to work instead of human labor in their factories, but it can also raise productivities for these employers.

- Can (AI) influence global economy growth?

Technological progress is main driver of growth of GDP per capita, allowing output to increase faster than labor and capital . However, technology can increase productivity, but also decrease the number of labor hours needed to create a unit of output. So (AI) causes unequal to labor wage decreases, even reduces the number of labor to manufacture, e.g. artificial intelligence technology of automation car manufacturing industry; clothing manufacturing industry; plane manufacturing etc. high technology of artificial intelligence manufacturing method. But (AI) should be potential environment benefit, although it raises unemployment ratio. Moreover, it can rise production , due to many skilled craft were replaced by the combination of machines and lower-skilled labor. The result of (AI) technology introduction , it causes output per hour risen when inequality declined, driving up average living standards, but the labor of some high-skill workers was no longer as valuable in the market. Otherwise, if (AI) technology is continue developed to be success. Some routine intensive occupations will be loss, which focused on predictable, e.g. easily programmable tasks, such as switchboard operators, filing clerks, travel agents, and assembly line workers would be particularly replaced by new (AI) technology. However, at the same time, (AI) technology development will bring these benefits: improvement in education (training (AI) technology scientists) , due to (AI) manufacturing technology needs are raising to businesses and institutional changes, such as the reduction in unionization and raising in the minimum wage to the (AI) manufacturing technology skilled labor in factories.

Because (AI) technology is not a single technology, but rather a collection of technologies that are applied to specific tasks, the effects of (AI) will be felt unevenly though the economy. It will bring some tasks will be most easily automated than others , and some jobs will be affected more than others, both negatively and positively. Finally, new jobs are likely to be directly created in areas , such as the development and supervision of (AI) as well as indirectly created in a range areas though out the economy as higher incomes lead to expanded demand.

However, if (AI) technology could dominate global labor markets. If labor productivity increases, do not influence into wage increases, then the large economic gains brought about by (AI) technology could be increased wealth inequality, due to employers can reduce production cost, but workers (labors) wages will not be increased, even will be decreased. Hence, it seems the (AI) technology will bring disadvantages to labor market to cause unemployment or reduce wages in possible, although it can reduce employer individual salary (wage) expenditure and it can raise productivity.

- How can artificial intelligence impact global economy growth?

Artificial intelligence (AI) technology is a branch of computer science that aims to create intelligent machines that work and react like humans. So, (AI) is a technology that appears to impact (influence) human preference by learning, understanding complex contents, enhancing humans in executing both routine and non-routine tasks. In the future, (AI) technology that can be virtual personal assistant, as well as it may exist, such as robots with human-like processing capabilities.

How can (AI) technology impact global economy growth over the next 10 years? During this time period, (AI) technology is predicted to have wide-ranging applications including: Machine learning that automates analytical model building by using algorithms that allow machines to operate without human assistance.

In global education aspect, potential applications include predicting cause-and-effect relationships from biological data, identifying new drugs, self-driving cars, and protecting against fraud, improved natural language processing that allows computers to continue to better analysis, understand and generate language to interface with humans using natural human languages. For example, transcribing notes dictated by physicians, automatically drafting articles and translating text and speech. So (AI) technology can be applied to education aspect to improve humans' knowledge level.

In visual art aspect, (AI) machine vision that allows computers to identify objects, scenes and activities in images. Current applications of (AI) machine vision include providing objective descriptions for the blind seeing(visual) needs.

We except the economic effects of (AI) technology to include both direct GDP growth from sectors that develop or manufacture. (AI) technology and indirect GDP growth through increased productivity in existing sectors that employ some form of (AI). If (AI) technology is an increasingly critical component of more products, it will become an integral part of many people's lives. Thus, (AI)'s ability to influence economic activity, rather than the economic or development status of the region. (AI) has the potential to impact income classes and to bring significant gains to both developed and developing countries. For example, (AI) has the potential to optimize good production around the world by analyzing agricultural regions and identifying what is necessary to improve crop yields.

In estimating the future economic effects by (AI) technology innovation, it is important to note that it is challenging to accurately predict which applications of (AI) will ultimately be commercially successful. In micro level economic influence, we need to apply methodologies to estimate the economic effects of investment in firms developing (AI) technology since investment levels in a technology are a telling sign of the future potential of that (AI) technology.

● How can (AI) influence GDP of high income countries in the next ten years?

How (AI)'s development may affect the global economy over the next ten years. In fact, (AI) technology has the potential to affect business across the global in a wide range of industries in ways only a number of technologies have done in the parts. For example, (AI) technology's expected to be a useful tool for enhancing human capabilities and in some instances replacing functions, such as driving a car, adoption of broadband internet, mobile telephone, industrial robotic automation have served to enhance human capabilities.

However, significant public debate has focused on projections of (AI) technology's effect on the labor force. However, large companies prefer to invest in (AI) technological industry. For example, face book's (AI) research lab., google machine intelligence lab. and micro soft machine learning and artificial intelligence research division are all making advances in (AI) technology and investing in the industry's top talent. Additionally, between 2010 year and 2015 year, nearly $5 billion in venture capital funding invested in firms across the global developing and employing (AI) technology (Facebook (AI) Research).

● How can artificial intelligence impact on workplace?

Modern information technologies and the labor economy growth of machines is powered by artificial intelligence have already strongly influenced the world of work in the 21 ST century. Computers, algorithms and software simplify every tasks and it is impossible to image how most of our life could be managed without them. How can be the information economy characterized by exponential growth replaces the most production industry based on economy of scales? What will the future world of work look like and how long will it take to get? Will the future world of work be a world where humans spend less time earning their livelihood? Alternatively, are mass unemployment, mass poverty and social distortions also possible scenario for the future, where robots, artificial intelligence systems play an increasingly central role? These questions concern how artificial intelligence further development . Can influence labor economy growth on workplace ? When the labor market has widespread impact on intelligence property, information technology, product liability, competition and labor and employment laws.

How (AI) technology impacts on labor workplace.

The future influence any organizations how labor economies use of (AI) can be analyzed, such as deep machine learning is based on a set of model high level data. Unlike human workers, the machines are connected the whole time in workplace. If one machine makes a mistake, all autonomous systems will keep this in mind and will avoid the same mistake the next time.

Over the long run intelligent machines will win against every human expert. Production robots have been replacing employees because of the (AI) technology. They work more precisely than humans and cost loss. Creative solutions like 3D printers and the self learning ability of these production robots will replace human workers, the automatic data recording and data processing, traditional back office activities are no longer in demand. Autonomous software will collect necessary information and will send it to the employee who needs it. Additionally, dematerialization leads to the phenomenon that traditional physical products are becoming software. For example, CD or DVDs are being replaced by streaming services. The replacement of traditional event ticket, e-travel ticket service products or hard cash will be the next step, due to the possibility of payment by smartphone. So, (AI) technology will impact human's daily life consumption behaviors in the future. For another example, transportation tools, such as boats and ferries and private vehicles will use sensors and navigating without human input. Taxi and truck drivers will become obsolete, the stock store applies to stock managers and postal carriers of the delivery is distributed by (AI) machine delivery method.

What is the relationship between (AI) and (CRM)?

- Can (AI) technology impact on customer relationship

management (CRM) ?
Nowadays , (AI) is a technology almost as old as the computer industry itself, it is similar with the advent of personal assistants function to businesses and personal promotion channel, such as (Amazon's Alexa, Apple's Siri, Google's Assistant) image recognition (face book), personalized recommendations (Netflix , Amazon). Those innovations have been driven by a increase in processing power, lower cost hardware, and the exploding creation and availability of data. It seems, (AI) technology can impact global customer service management method.
How to forecast economic impact modeling to (AI) will affect global economy? Can human forecast business revenue growth and job creation (or destruction) based on (AI) applied to customer relationship management (CRM) activities? In addition to the economic impact on (AI) or (CRM) which can include an estimate of the economic impact attributable to sales forces customer base. What can economic benefits be brought to (CRM) from (AI) technology?
Artificial intelligence(AI) comprises a set of technologies that use natural language processing, machine learning, knowledge graphs, and other tools to answer questions, discover insights and provide recommendations. Computer systems can use (AI) hypothesize and formulate possible answers based on available evidence can be trained through the ingestion of vast amounts of content, and automatically adapt and learn from (AI) self mistakes and failures.
So, any business organizations (customer service departments) can provide efficient and effective customer relationship management of excellent customer service quality if which applied (AI) technology system. The different type of (AI) systems include: (AI) system platforms, machine learning (AI) based data preparation and enrichment tools, machine vision/image recognition, voice speech recognition, text analysis and natural language processing, bots , e.g. face book website and virtual digital assistance solutions, social media pattern analysis , sentiment analysis, advanced numerical analysis (e.g. IOT streaming , machine logs), supporting

technologies, knowledge base dialog management, Q&A processing etc. different (AI) technology system customer relationship management (CRM) tools.

(AI) (CRM) of activity can include these categories, such as: corporate marketing, marketing operation, field marketing, customer support, digital commerce, customer analytics, customer influenced product or service design, product or service pricing, finance information, presentation, customer billing, inventory , logistics and fulfilment support, partner management etc. different CRM tools.

(AI) technology of CRM has been carrying on plan different stages to achieve CRM personal assistant tool for businesses. The stages are such as, in the beginning stage of (AI) projects in place, implement now, pilot phase next year in the final stage of (AI) customer relationship management tools are foreseeable future. So, this CRM technology has been improved to plan in different stages every year to prepare to achieve full capacity of CRM service quality for businesses to use in the future.

Hence, how to develop an estimate prediction of the economic impact (AI) technologies could have CRM activities, which depends on gathering macroeconomic information on business revenue and the basic marketing of business revenue and the basic markup of business expenses by major functions (customer support, marketing and sales , production etc.)

An economic impact model that can gather data together and forecast the results how (AI) artificial intelligence technology brings (CRM) customer relationship management benefits to businesses, e.g. surveys investigation includes IT spending by sample countries, GDP and population estimates and forecasts, revenue per employee and ratios of IT spend to GDP. Surveys (questionnaire questions) of forecast results are influenced by (AI) impact can include: results are projected from surveys and rely on estimates are made by respondents on the expected financial improvements in categories of (AI) –assisted customer relationship management activities. The forecast assumes that these estimates are correct; financial estimates are based on estimates of "first year" improvement from full (AI) implementation; forecasts are from planning to implement any artificial intelligence of customer relationship management (CRM) projects, the improvement forecast is of categories of activity , e.g. corporate marketing , digital commerce, and customer analytics. They are not estimates of ROI for the (AI) software. They rely on conservative estimates to which each of these entities might affect company revenue, expenses or productivity. They also rely on estimates of the penetration of software in customer relationship management activities . Net new jobs created are based on the ratio of new revenue to jobs required to support that revenue . They can assume that 50% of the net new revenue will support increases in labor and the rest will go for capital and other operating expenses that may replace jobs lost to automation.

In the future, some of the ways in micro economic benefits to any organizations. (AI) technology is expected to impact CRM activities include: Spending up sales cycles, improving lead generation and qualification solving customer support problems faster (raising service quality), helping companies improve brand campaigns and recognition, lowering costs of support calls when increasing resolution rates, lowering the cost of recruiting employees and partners, increasing revenue from optimized product marketing, optimizing price, distribution logistics and preventing loss through fraud detection. So, micro economic benefits view point, it seems that (AI) CRM technology can raise any companies economic benefits for care term.

Artificial intelligence enables machines or the in-build software to behave like human beings which allows these decisions and act. The advent of (AI) is leading , talking, making decisions and act. The advent of (AI) is leading to new technologies advances and transforming the economic and employment opportunities for humans in a positive way. (AI) related technologies can facilitate our live. For example, industrial robotics, robotic medical assistants, smart games, financial forecasting software, big data analysis, algorithms in health and bioinformatics, pilotless cargo places,

drone ambulances and general purpose and workplace robots and others. (Disruptors technologies: Advances that will transform life, business and the global economy).
Artificial intelligence also known as computational intelligence is defined as " the human –like intelligence exhibited by machines or software. It is theorized that intelligence of humans can be described and intelligence machines or software can simulate it. These machines software can be reasonable , learn, perceive and process information, like human mind and thus facilitate human life. They can think and act for us. So, artificial intelligence is an interdisciplinary field of study including computer science, neuroscience, psychology, linguistics and philosophy.
However, (AI) research and developments have economically impacted many industries, such as robotics, telecommunications, computer applications , health, finance, heavy manufacturing, transportation, aviation, e-service and e-commerce, military , music and movie, toys and games entertainment etc. industries.
In fact, many ideas, systems and technologies have been developing in the world of (AI) technology. However, which are net called or considered (AI) products, rather which are mentioned with their specific names, such as smart graphics, machine learning, e-commerce etc. (i.e. this is called (AI) effect).

- How can (AI) technology influence digital economy?

Nowadays, (AI) related industrial applications will replace most human power in fields, including call centers, customer services and air cargo transportation. (AI) technologies also help weather forecasting based on repeated rainfall pattern (data) recognition, through robotics (i.e. floor cleaning, moving lawns etc.) transporting people and products with unmanned vehicles, sending space unmanned smart shuttles, developing robotic arms, predicting market values in stock exchanges by internet, making homes safer, helping elderly and disabled using robotic servants etc.

Among the (AI) related technologies , there are a few that

significance for the impact on society and especially on digital economy . (AI) is particularly influential in machine learning. Such as robotics, transportation, finance, health and bioinformatics, e-commerce , e-games, big online data gathering and internet-of-things. For example, machine e-learning is based in bioinformatics and robots that can learn new skills for better caregiving in healthcare. What is machine e-learning? Machines can e-learn from e-data gathering, coming up generalizations and making decisions to act in certain ways from internet.

There are important applications , such as e-machine perception, electronic online natural language learning processing, online search engines, online bioinformatics, online brain –computer interface, online game playing, online robot locomotion, online advertising, online computations finances, online health monitoring, online DNA classification and decision making, online in chemistry –cheminformatics . So, online machine learning can positively impact productivity and it can enhance information and analytical system from (AI) online channel.

What is robotics? Robotics is one of the most strongly influenced fields in (AI). For example, heavy manufacturing industries, robots and used and man power is replaced for effectiveness, precision, and accuracy, especially in respective or dangerous tasks, including welding, assembling , picking and placing .

So, robots can acquire new skills or adapt the changing dynamic environment. Also, artificial intelligence can be applied in developing transportation. For example, automated vehicles, driver assistance systems , safety systems, collision avoidance systems and public transportation. Moreover, (AI) technology has proven to produce some of the best tools to predict stock market fluctuations from internet data gathering method. It's predictions are based on ever-evolving predictions algorithms and systems learn new models and make connections between historical data and new data to measure stock market trading more accurate from internet data gathering channel.

In health field, especially in health data processing , analysis, decision making support and medical diagnosis. So, online data can show which patients will need what treatment and what alternative drugs could be used more accurate from (AI) online data gathering method. Bioinformatics is an interdisciplinary field combining statistics, (AI) online technology can help in discovering data patterns and modeling through the application of machine learning, artificial neural networks and genetic algorithms. For example, further (AI) technology development of human genome project of online data sequences.

Online shopping can be facilitated by virtual assistants developed through (AI) technology and these assistants can offer the best advice. (AI) online purchase coming after every product image recommendations and personalization bring important revenue to shopping online sites, like Amazon . Smart computer graphics and games, artificial intelligence is useful in smarter computer, graphics, scene modeling , scene rendering processes in order to create, for example, effective human –robot interactions , online machine learning, online strategic games techniques etc. online computer related (AI) software.

So, online big data analysis and big data does have a critical need in the world of online intelligence machines and software in our future. In other words, (AI) offers online technology to enable online big data analysis to provide industrial organizations with valuable information for effective decision making in short time. For example, what IBM's Watson achieved: this machine used 200 million of structured and unstructured content with a special technology of hypothesis generation, massive evidence gathering, analysis and scoring from internet channel.

Finally, (AI) online technology another related internet invention (internet of things) (IOT) is the network of machines or objects connected through internet. These connected objects can sense their internal and external environment, communicate with each other, can send critical data and finally can make decisions to act or correct their environment from (AI) online technology. For example, factories can monitor and automatically change production

processes, hospitals can monitor and regulate the health conditions of their patients , schools can collect data from facilities and cars can send data to car makers from (AI) online technology.
Partner predicts that (IOT) market will create about trillion amount value by 2020 year. Although machines collect big data from their environment, whether which gain an insight or learn from these online data largely depends on the (AI) online machine learning principals and (AI) online technology. In 2013, Mckinsey estimated that disruptive technologies closely related with potential economic impact in 2025 year between $7.1 to $13.1 trillion amount (automation of knowledge work, advanced robotics, autonomous or near-autonomous vehicles).

What is the relationship between
(AI) and global digital economy development ?

● Could work activities in China be automated
making in the nation with the world's largest automation potential? Can (AI) technology influence China economy? Could China workers be affected and jobs made up of routine work activities and predictable? Will programmable tasks be particularly impact to China employment market ? When impact on labor market is likely to be gradual at the aggregate level, it can be sudden and dramatic at the level of specific work activities, rending some job obsolete fairly. Overall (AI) technology will raise digital skills when reducing demand for medium incomer inequality for China workers. It seems (AI) technology's effect on productivity could be crucial to China's future economic growth as the population ages are increasing.
In China, some biggest technological companies driving significant investments in research and development. Moreover, China is one of the leading global (AI) technology development county. However, China will need to focus on building its innovation capacity. For example, United States and United Kingdom are currently producing more influential (AI) technological research. However, if China planed to achieve (AI) technology success, it's traditional

industries will need to develop technical know-how –to and overcoming implementation costs prepare to develop (AI) . When (AI) technology is introduced into China society, China government needs to raise concerning ethical, legal, technological security etc. business questions. Also, surrounding issues include privacy, discrimination, legal liability and regulation. It aims to encourage overseas investors to choose to invest (AI) technological industry to raise GDP growth and manufacturing industries income growth for long term in China.

If China encouraged overseas (AI) technology investment in its country. It is possible to influence China employment market to be changed. Because (AI) technology will impact to influence China people daily life. Due to (AI) technology is introduced to China society, many rich people will prefer to spend to buy any high (AI) technological products for entertainment or learning or machine man driving etc. daily necessity activities. Then it will raise GDP growth and will raise (AI) manufacturers or related-(AI) technological manufacturers profit. It is beneficial to China because it can become one high knowledgeable and (AI) technological economical society.

But it will bring bad influences to raise unemployment chance for the low skillful labor. In labor economy aspect influence , how (AI) technology can influence China low skillful labor unemployment ratio raising. The raising low skill labor unemployment reason is because China low skillful human labors are argued or are replaced by (AI) technology creating new challenges to introduce to influence China society of simply human manufacturing job nature to be changed to be high (AI) technology manufacturing job nature in any China factories. Moreover, when (AI) technology introduction to China, it will cause other related social challenges in China. The varied (AI) related challenges, including the difficulty of creating safe and reliable hardware for sensing and affecting (transportation and education), the challenges of gaining public trust, a low resource comities and public safety and security, the challenges of overcoming fears or marginalizing humans in China employment and workplace and the risk of diminishing interpersonal trust because the low skillful labors won't believe any China employers will give chance to employ them , due to (AI) technology will replace their skills and man manufacturing of productivity is much less to compare to (AI) technology manufacturing method.

- How does (AI) technology influence

the future of employment change?

Are future nature of jobs changed to computerization from (AI) technology? Where are the probability of computing occupations from (AI) technology influence? What is expected impacts of future computing on labor market from (AI) technology influence? John Maynard Keynes's frequently cited prediction of widespread technological unemployment " du to our discovery of means of economic the use of labor outrunning the pace of which we can find new used of labor" (Keynes, 1933, p.3).

In the future, (AI) technology will impact some nature of occupations to change computing. This chance will also influence some countries' economic change. For example, some factory human labors hand routine manufacturing tasks will be changed to computerization of routine manufacturing tasks by (AI) technological machine men hand manufacturing method. it will cause a structured shift in the labor market, with workers reallocating their labor supply from middle-income manufacturing to low-income service occupations.

Arguably, this is because the manual tasks of service occupations are less computerization, as who require a higher degree of flexibility and physical adaptability. So, (AI) technology will influence the human hand labor skillful occupation nature of task cheaper , such as vehicle manufacturing , ship manufacturing, computer manufacturing, steel manufacturing, television, radio etc. home electronic products of heavy machine industry change. Due to (AI) technology machine man will be proper to be used to manufacturing these electronic products when the (AI) technology innovation can develop to the mature stage. Then, any countries

manufacturers will choose to use (AI) technology machine man, instead of human hand production.
Supposing the future prices of computing are fallen, seriously, problem solving skills are becoming relatively productive, explaining the substantial employment growth in manufacturing occupations, involving cognitive tasks where skilled labor has a comparative advantage, as well as the increase education needs for (AI) technology computing of machine man subject study.
Prediction of education needs for (AI) technology student numbers will increase, due to manufacturing industry needs many (AI) technology students in future employment market. Another (AI) technology influence if the future (AI) technological innovation, e.g. machine man manufacturing or machine man service industries will both increase demand, then with more sophistic software technologies will be disrupted labor markets by marketing workers redundant.
For publishing industry, what is striking about the case in paper book publishing industry will be unpopular? Due to the electronic book publishing industry will be popular, e.g. Amazon publish . (AI) technology can influence paper book manufacturing method which is replaced by machine man electronic book manufacturing method as well as it will cause the computerization is no longer confined to routine manufacturing tasks. Due to (AI) machine man manufacturing technology will be proper to be used to manufacture any products in short time efficiently and effectively , e.g. electronic book products. In the future, if it is fact to occur this case, such as (AI) technological machine man manufacturing method will be adopted (applied) to manufacture electronic books or any products in possible. (AI) technology will cause many manufacturing workers are unemployed. It is beneficial to employers, who can reduce to spend much wages expenditure to employ manufacturing workers, but it will cause many manufacturing workers loss jobs and reduce income to support whose families lives. It will cause social challenges, e.g. increasing stealing crimes if the manufacturing workers had not other skills to find other jobs to do easily. So,

manufacturers need to concern over technological unemployment which will be hardly future phenomenon if who decided to dismiss all manufacturing workers, due to (AI) technology machine men replace to them.

If (AI) technology can be innovated to produce any kinds of machine man to serve any service or manufacturing industries successfully. Then, it will bring these questions: Can future that workers be influenced to be automation employment and productivity by (AI) technology influence? Does it impact to influence the (AI) technology countries' productivity and growth and natural resources development and labor markets and evolution of global financial markets and economic impact of technology and innovation and urbanization etc. issues? How will automation transform the workplace? What will be the implication for employment? What is likely to be its impact both on productivity in the global economy and on employment?

In fact, automatic of activities can enable businesses to improve performance by reducing errors chance and improving quality and speed, and same cases achieving outcomes that go beyond human capabilities. Some economists indicate (AI) technology would give a needed boost to economic growth and prosperity have of the working age population in many countries. Based on the scenario modeling, they estimate automation could raise productivity growth globally by 0.8 to 1.4 % annually. They also indicated that almost half the activities people are almost $1.6 trillion in wages to do in the global economy have the potential to be automated adapting current demonstrates technology, according to their analysis of more than 2,000 work activities across 800 occupations. When less than 5% of all occupations can be automated entirely using demonstrated technology, about 60% of all occupations have at least 30% of worker made activities, that would be automated. More occupation will change to be automated. They also indicated for business performance benefits of automation are relatively clear, but the issues are more complicated by policy making to attract foreign investors. Beyond technical feasibility, the cost of technology, competition labor will include skills and supply and demand dynamics, performance benefits and beyond labor cost savings and social and regulatory acceptance will affect the automation. Their predictions suggest that half of today work activities could be automated by 2055 year, but this could happen 10 to 20 years earlier or latter depending on the various factors in addition to their wider economic condition.

Some scientists suggest (AI) technology is finally starting to deliver real-life business benefits. Computer power is growing significantly , algorithms are becoming more sophisticated and perhaps most important of all, the world is generating vast quantities of the fuel that powers (AI) technology data billions of gigabytes of it every day. Also, online firms are digital natives, such as Google online search service company is investing on (AI) technology. For new though most of the news if coming from the suppliers of (AI) technologies. And many new users are only in the experimental phase. Few products are on the market or are likely to arrive these soon to drive immediate and widespread adoption. As a result, analysts believe (AI) technology's potential will give true economic benefit in the future. (AI) industry will introduce to suppliers and users to raise economic potential of (AI) technology.

In the future, (AI) technology systems can solve business problems. Some scientists categorized those into five technology systems that are key areas of (AI) technology development: robotics and autonomous vehicles, computer vision language virtual agents and machine learning , which is based on algorithms that learn from data without replying on rules-based programming in order to draw conclusions or direct an action.

Such as computer vision and language includes natural language processing, analytics, speech recognition technology, some are about learning from information, such as about machine learning and others are related to acting on information, such as robotics, autonomous vehicles and virtual agents, which are computer programs that can converse with humans. Machine learning and a subfield called deep learning are artificial intelligence applications.

- Can artificial intelligence impact global economy growth?

Artificial intelligence (AI) is a term first defined in 1956 year. It is a branch of computer science that aims to create intelligent machines that work and react like humans. In contrast today, 60 years later, (AI) is characterized by a number of applications, including computers playing games against humans and understanding human languages, virtual personal assistants, and robotics which involve computers seeing , hearing and reacting to sensory stimuli. In the future, technologists predict for (AI) technology ranging from (AI) being used as a tool to aid relatively simple processes for robots with human like mental capabilities, who expect (AI) technology can emulate human performance by learning, coming to mind its own conclusions, understanding complex content, engaging in dialog with people, enhancing human cognitive performance or replacing humans in executing both routine and non-routine tasks. In existing industry, (AI) technology is used , such as targeted advertising and virtual used personal assistant as well as the (AI) technology that my exist in the future, such as robots with human vehicle processing capabilities.

The range of (AI) technology's progress in the future will determine the economic impact future of (AI) technology on the global economy with more limited advances and applications (i.e. weak (AI) only) corresponding to more limited economic impacts and more substantial progress, i.e. strong (AI) technology is corresponding to more significant economic impact.

(AI) technology learning that automates analytical model, including predicting cause-and-effect relationship from biological data, identifying new drugs, self-driving cars and protecting against fraud etc. functions. Also (AI) learning can improve natural language processing that allows computers to continue to better analyze, understand and generate language to interface with human using the natural human language, virtual personal assistant, helps users by providing scheduling appointment, reminds organizing personal finance and finding providers of

various services, machine vision allows (AI) machine man to identify object, scenes and activities in detect pedestrians and bicyclists.
We expect the economic effects of (AI) technology to include both direct GDP growth from sectors that develop or manufacture (AI) technology and indirect GDP growth through increased productivity in existing sectors that employ some from of (AI) technology. If (AI) producing sectors could grow, then it could lead to increase revenues and employment of (AI) technological professionals within these existing firms as well as the potential creation of entirely new economic activities to any countries' societies productivity improvement in existing sectors could be realized through faster and move efficient processes and decision making as well as increased (AI) technological knowledge and access to information available in societies easily.
In the future, if (AI) technology is an increasingly critical component of more products, it will become an integral part of necessary products of many people's lives. The extent of (AI)'s economy effort is also likely to vary from region to region, thought variation may be more dependent on the predominate economic activity of a region and the (AI) ability can influence economic activity, rather then the economic or developmental status of the regions. (AI) technology can move accessibility and can use source development to do international business between one country and another country.
So (AI) technology has the potential to give benefits to different income chooses and to bring significant gains to both developed and developing countries. For agricultural technology, (AI) has the potential to optimize food production around the world by analyzing agricultural regions and identifying what is necessary to improve crop yield. In total, (AI) technology gives greater economic impact to any countries agricultural regions if which implemented (AI) technology to grow crop , fruit etc. food production in the farms. Investment in (AI) technology is such as capital investment to any countries' public or private enterprises. So, it will have large

economic impact to the future . If the (AI) technology is reasonable invested to the different needs aspect by the public or private enterprises in the country. Then, it will have good economic impact to the country in the future. However, when (AI) technology is likely to affect both the productivity and employment components of economic growth in many sectors. Significant public debate has focused on projections of (AI)'s effect on the labor force. However, for instance, some researchers have argued that the rise of (AI) technology and automation will led to significant unemployment as capital is substituted for the low skillful labor. So, they point to the concern that the increasing sophistication of (AI) technology may balance skilled and semi-skilled workers and the reduce the size of the middle class. However, this is not a new argument, due to (AI) technology negatively affecting the labor force and leading to mass unemployment. Because the (AI) technology is the substitution of machinery for human labor. Although, employment in certain industries, has been reduced in the past due to technological advancement. For long term, the labor market has adapted to the introduction of new technology, giving rise to new jobs in new areas. (AI) technology may also be accomplished without a reduction to total employment in the long-term to some Asia countries, such as Hong Kong and Japan. Because Hong Kong and Japan many low skilled labor, e.g. security, cleaner who complaint that employers need them to work long time hours. (abnormal working hours) e.g. one day 12 to 15 working hour per day. Hence, if (AI) machine means invention technology success. Security or cleaning job can be worked from (AI) machine man in some hours every day in order to reduce the long time working hours cleaners or security workers, e.g. one (AI) machine man works 4 hours for cleaning or security job, one day as well as another cleaner or security labor only needs to work 8 hours one day. So total security or cleaning employers can employ 12 hours machine cleaners or security workers and human cleaners or security workers in one day. For long term benefit, Hong Kong or Japan every security or cleaning worker does not need to work 12 hours minimum working hours one day. They won't feel tried and bore and without private with whose families, so who will accept to do these cleaning or security jobs, even they can raise work efficient and performance when who feel happy and health.

So, (AI) technology of machine man invention can raise low skillful labor efficiency and it can help them to avoid abnormal working hours demand in some busy work life countries, such as Hong Kong and Japan. Before, one Japan female labor feel unhappy to work, due to who often needs to work abnormal working hours for her employer and who has less sleeping and without any private time to enjoy her life with her families every day. So this abnormal working hours factor causes her to do commit suicide behavior, then she is die unlucky. So (AI) technology of machine man invention ought avoid abnormal working hours demand for employer in any countries in the future.

The most important occurrence to any employers, some researchers had attempted to do one experiment to find that private research and development , venture capital and public research and development investment all have strong net effect or economic growth with venture capital funding further having the strongest such effect from (AI) technology. The researchers hypothesize the venture capital investment contributes to economic growth through (AI) technology innovation and by the capacity of an economy to use existing (AI) technology knowledge to increase productivity. They predict the impacts of venture capital, business-research and development and public research and development can raise multi factor productivity from (AI) technology introduction.

Can (AI) technology influence the economic development to developing countries? The developing regions of the world contain most of natural resources. If one day, (AI) technology has invent one kind of machine man which can assist any gas or oil workers to seek any new oil/gas natural resource locations easily. I believe that (AI) technology can help these natural resource exploitation countries will gain economic benefit more easily. So, (AI) driven technology can be used to change to create any new opportunities to address poor management or resources and improve human well being, such as Africa Latin America and India can use (AI) technology machine man to seek any oil/gas natural resource countries exploitation activities to attempt to gain much economic benefits.

- Why will (AI) technology grow economic development ?

Nowadays, increases in capital and labor are no longer driving the levels of economic growth, such as (AI) technology. The ability of increase in capital investment and in labor of traditional drivers of production, have no longer to be enjoyed in most developed economies ,e.g. developed country, US, UK . However, artificial intelligence has the potential to overcome the physical limitation of capital and labor to avoid missing out on this opportunity. So, policy makers and business leaders must prepare for and work toward a future with artificial intelligence. They must do with the idea that (AI) is another simply method to enhance productivity method . Rather they must see (AI) as the tool that can transform thinking about how growth is created.

Economists have always thought of new technologies are as driving growth their ability to enhancing. It can replace labor and capital factor of production. So, it brings this question: What is the factor of production (AI) technology characteristics. They key factor is to see (AI) technology as a capital-labor .

(AI) can replicate labor activities at much greater scale and speed, and to even perform some tasks began the capabilities of human. For example, by using virtual assistants , 1000 legal documents can be reviewed in a matter of days instead of taking three people six moths to complete. Some (AI) technology may be one kind of factor of production in the future. For another example, people will work in workplace digitalization environment. So, in the future, working environment and information management are automated. Such as Konica camera sale company will use workplace digitalization. So , (AI) technology can provide workplace digitalization in order to raise productivity efficiency. (AI) technology will be one kind of production which is replaced by workplace digitalization and it will grow any organization productivity efficiently. Then, (AI) technology will assist overall social economy growth , due to productivity is raised and products can be produced in short time to prepare to sell in consumption market. So, time will be shortened to increase GDP growth fast for the development of (AI) technology countries.

● How can (AI) technology impact to global economic and social and psychological changes?

What will be the development of (AI) technology and predictions concerning the future evolution? The computers and robots will develop conscious, intelligent and minds into humans, enhancing psychological and behavioral abilities and allowing for direct communication with (AI) minds. (AI) technology will be impacted human life by (AI) technology information communicative and environmental influence. A " world brain" and " world mind", this psychological system will be enhanced and enriched the capacities of both individual and collective cognition by (AI) technology of service industries.

(AI) technology with influence these human needs of service industries changes, such as , biological science, finance, entertainment, business, biological science, transportation, communication military etc. The personal computer evolution, the internet and the world wide web which exploded on the scene, linking business, homes, schools, social organizations which were a completely unpredicted phenomenon to influence human life. Kurzweil (1999) predicts that by 2029 year, most human communication will be with machines. According to Person, by 2100 year, there will be human machine convergence.

How can (AI) technology influence environmental protection to make benefits to farming economic growth? (AI) technology can be applied to predict how to solve environmental pollution challenge to avoid to damage any crop or vegetable or rice or fruit etc. food growth. Because environmental experts can gather global environmental pollution data from an environmental database to build a perform a systematic analysis from (AI) technology. The first step is this broad analysis can include understanding, statistical and data gathering techniques to obtain the relevant data, the correlation among the variables involved, and a list of possible models. The next step is to select a set of methods and models that cover all kinds of knowledge and functionalities needed for the decision making process. Once the models are selected, they must be fully implemented by means of machine learning , data mining, statistical or numerical technique. After that, those models must be integrated to build the whole EDSS. The EDSS must be tested to check its performance, accuracy, usefulness and reliability, both from the user's and (AI) technology/computer scientist's point of view. If these is any wrong feature in any development stage, such as model's integration, models' implementation, selection of models, database, problem analysis etc. the developers must come back in the update th required components. When the evaluation phase is all right, the EDSS is ready to be applied to the environment. The great contribution of artificial intelligence to EDSS the integration of several methods complementing the classical statistical models/ simulation , statistical analysis, linear models, etc. and numerical models (control algorithms, optimization techniques etc.) .

This cooperation makes the resulting systems more reliable and powerful in coping with real world environment systems. Date interpretation has been a principal area of research in (AI) technology since the very beginning. The most demanding problem in the environmental assessment context. Knowledge representation permits the definition of the different types of data that the existing methods adapt to the process. There is also a lot of work to clean, repair and transform the huge available quantities of raw data. Apart from this, the availability of meta-information or background knowledge is required to guide the process. Data mining is multi-disciplinary: It covers expert systems, data based technology, statistics, data visualization and unsupervised machine learning. These techniques operate at the level of data and background information, where numerous and often incompatible new commensurate pieces of information from disparate sources have to be brought together (K, Fedra, 1994).

So, it seems that in the future, (AI) technology with the increasing maturity in particular those related to knowledge and engineering, new dimensions can be assisted to users in environmental decision making are available. For example, many environmental systems are characterized both by incomplete models and by limited data. Hence, in the future, (AI) technology will be applied to predict climate change to reduce crop or fruit etc. food agriculture challenge by climate change bad influence.

- Will (AI) technology influence digital economy change to manufacturing industry ?

To understand how the manufacturing business must adapt to prosper in the technology, we need to understand how (AI) technology will change us to shape our daily habits to satisfy our expectation of products to how we shop and even the immediate of the entire process. For example, taxi services are in the crosshairs as on demand transportation services like, available of the touch of a smart phone button expand. In fact, Yellow lab, US country , san Francisco city's largest taxi company is filing for bankruptcy as the industry starts to change faster than almost anyone expected. However, at this point, its more than an app that is changing, some our taxi passengers renting taxi transportation to catch consumption behavior.

(AI) technology will influence digital economy for taxi passenger's individual customer experience, offering a growing renting taxi to catch of service and feedback opportunities when any one taxi passenger who chooses to use mobile phone app online tool to prepaid to rent any taxi more easily.

Also in the long term, (AI) technology can influence vehicles drive themselves of behavior. Already, companies like Google and GM are working on projects to bring fleets of autonomous vehicles

to cities at the path of a button.
Moreover, this on-demand service model is beginning to appear across a much broader range of markets. For example , Amazon company is investing in its own fleet of trucks, planes and even drone at the same time as it pushes for same-day delivery of products. As some point, vehicles will be autonomous too. So, it seems that (AI) technique will influence any transportations choose to use digital autonomous driving technology in the future . For Amazon company case, it is not stopping of logistics. It is also aiming to automatically manage the supply of consumer home products with its recently launched Amazon replenishment service, Dash. Dash is a digital service that enables that connected derive to automatically order physical products from Amazon when supplies are running low. So, it seems (AI) technology will be applied to logistic function by digital technology method introduction in the future.

Hence autonomous vehicles will optimize industry supply chains and logistics operations through increased efficiency and flexibility. In fact, fully automated and lean supply chains will keep reduce load sizes and inventory by leveraging smart distribution technologies and smaller autonomous vehicles by machine man assistance. If Amazon continues to grow market share for online sales by reducing effort required by the consumer to place an order, when also contributing the almost immediate delivery of products to the doorstep. So, it will further fuel the trend toward on-demand derive. As Amazon company fuels the on-demand economy, consumers will expect immediacy in more parts of the digital economy. On top of speed, consumers increasing expect more personalization options.
So, (AI) technology will influence digital manufacturing, such as Amazon publishing to monitor every aspect of every process in real -time and communicating to self-optimized deep learning robotics, new methods of high volume and high customization will become possible. Then, as products merge into product platforms and even

services, manufacturers have the opportunity to provide components and platforms used by smaller players. So, (AI) technology will influence manufacturing industry to choose automated SMI lines, robots installed, automation engineers.

Another future (AI) technology development can be applied to space science aspect, such as Automation engineering space in manufacturing process to achieve digital manufacturing benefits to any businesses in the future. Such as reducing cost, shortening manufacturing time, raising efficiency, shortening delivery products to client individual time. How can artificial intelligence give the need and advanced fast and evaluation methods benefits for space exploration? When US NASA (space exploration organization) achieves any space exploration missions, it will answer this question:
When is it useful to have a machine use (AI) technology to achieve a decision? After all, after millions of years of space exploration and rough 10,000 years of civilization, humans are usually quite good at making decisions in complex uncertain environments. Through, Johns Hoplains University's Applied Physical Lab. Research in (AI) technology enabled systems, which has identified three general use cases for (AI) technology to explore space mission:
First, for some tasks (AI) technology is more cost effectiveness than human. Second, (AI) technology is better suited than humans at solving some, but not all problems. Third, (AI) technology allows NASA organization's space exploration mission to develop machines that ate capable of responding faster than when a human is in the decision loop (D. Scheidt, 2012, A. Castano et. al. 2008).

So, the use of (AI) technology to enable science by observing the pace of rapidly evolving phenomena was demonstrated. It is more effectively coordinating and (AI) technology utilizing to earn economic benefits to use for space exploration mission.
However, (AI) technology also have current risk for space exploration. Today (AI) technology is immature and requires further development to reach its potential. For instance, the (AI) technology algorithms that detected the dust derive could not have

identified whether the Martain weather represented a threat to the cover. Also it can not yet use instrument input to determine what, where and how to autonomously make the next space science measurement. An equally important factor limiting (AI)'s deployment is that lacks the methodology and technology to effectively test (AI) technology. So, the challenge will testing (AI) enabled system is how (AI) performance can be measured. It would be NASA organization's difficulty to find (AI) technology to develop to carry on researching any space exploration missions in the future. However, (AI) technology will be a good economic benefit choice for space exploration mission in the future.

chooses to use mobile phone app online tool to prepaid to rent any taxi more easily.

Also in the long term, (AI) technology can influence vehicles drive themselves of behavior. Already, companies like Google and GM are working on projects to bring fleets of autonomous vehicles to cities at the path of a button.
Moreover, this on-demand service model is beginning to appear across a much broader range of markets. For example , Amazon company is investing in its own fleet of trucks, planes and even drone at the same time as it pushes for same-day delivery of products. As some point, vehicles will be autonomous too. So, it seems that (AI) technique will influence any transportations choose to use digital autonomous driving technology in the future . For Amazon company case, it is not stopping of logistics. It is also aiming to automatically manage the supply of consumer home products with its recently launched Amazon replenishment service, Dash. Dash is a digital service that enables that connected derive to automatically order physical products from Amazon when supplies are running low. So, it seems (AI) technology will be applied to logistic function by digital technology method introduction in the future.

Hence autonomous vehicles will optimize industry supply chains

and logistics operations through increased efficiency and flexibility. In fact, fully automated and lean supply chains will keep reduce load sizes and inventory by leveraging smart distribution technologies and smaller autonomous vehicles by machine man assistance. If Amazon continues to grow market share for online sales by reducing effort required by the consumer to place an order, when also contributing the almost immediate delivery of products to the doorstep. So, it will further fuel the trend toward on-demand derive. As Amazon company fuels the on-demand economy, consumers will expect immediacy in more parts of the digital economy. On top of speed, consumers increasing expect more personalization options.
So, (AI) technology will influence digital manufacturing, such as Amazon publishing to monitor every aspect of every process in real -time and communicating to self-optimized deep learning robotics, new methods of high volume and high customization will become possible. Then, as products merge into product platforms and even services, manufacturers have the opportunity to provide components and platforms used by smaller players. So, (AI) technology will influence manufacturing industry to choose automated SMI lines, robots installed, automation engineers.

Another future (AI) technology development can be applied to space science aspect, such as Automation engineering space in manufacturing process to achieve digital manufacturing benefits to any businesses in the future. Such as reducing cost, shortening manufacturing time, raising efficiency, shortening delivery products to client individual time. How can artificial intelligence give the need and advanced fast and evaluation methods benefits for space exploration? When US NASA (space exploration organization) achieves any space exploration missions, it will answer this question:
When is it useful to have a machine use (AI) technology to achieve a decision? After all, after millions of years of space exploration and rough 10,000 years of civilization, humans are usually quite good at making decisions in complex uncertain environments. Through,

Johns Hoplains University's Applied Physical Lab. Research in (AI) technology enabled systems, which has identified three general use cases for (AI) technology to explore space mission:
First, for some tasks (AI) technology is more cost effectiveness than human. Second, (AI) technology is better suited than humans at solving some, but not all problems. Third, (AI) technology allows NASA organization's space exploration mission to develop machines that ate capable of responding faster than when a human is in the decision loop (D. Scheidt, 2012, A. Castano et. al. 2008).

So, the use of (AI) technology to enable science by observing the pace of rapidly evolving phenomena was demonstrated. It is more effectively coordinating and (AI) technology utilizing to earn economic benefits to use for space exploration mission.
However, (AI) technology also have current risk for space exploration. Today (AI) technology is immature and requires further development to reach its potential. For instance, the (AI) technology algorithms that detected the dust derive could not have identified whether the Martain weather represented a threat to the cover. Also it can not yet use instrument input to determine what, where and how to autonomously make the next space science measurement. An equally important factor limiting (AI)'s deployment is that lacks the methodology and technology to effectively test (AI) technology. So, the challenge will testing (AI) enabled system is how (AI) performance can be measured. It would be NASA organization's difficulty to find (AI) technology to develop to carry on researching any space exploration missions in the future. However, (AI) technology will be a good economic benefit choice for space exploration mission in the future.

- What is artificial intelligence potential
benefits and ethical considerations?

The ability of (AI) technology systems to transform vast amounts of complex information into insight has the potential to help solve manufacturing or service challenges for human needs. However, to reap the societal benefits of (AI) systems, humans will need to trust

then and make sure that which follow the same ethical principles, moral values, professional codes and social norms that we humans would follow in the same scenario, research and educational efforts as well as carefully designed regulation in order to achieve the most effort of economic benefits goals. For example, international business machines corporation (IBM) is actively engaged both competitors , in global discussions about how to make (AI) ethical and as beneficial as possible for people as social economic benefits.
(AI) is usually defined as the " capability of a computer program to perform tasks or reasoning processes " that human usually associate to intelligence in a human being. Often, it has to do with the ability to make a good decision, even when there is uncertainty, too much information to handle. As an example, play chess or complex card games of entertainment activities is believed to need some form of intelligence in a human being, as well as choosing the best medical facilities in a difficult medical case, or creating something new, such as mathematical theorem or even some form of act, or even driving automatic machine man (self driving vehicle) replacing human driving in the middle of a crowded city.
(AI) needs depends on what we consider being intelligence in the behavior of a human being act a certain point in time. If human belief about human intelligence changes and we don't believe any longer that a certain task requires intelligence, then a computer program performing that task is no longer part of (AI), it becomes just another boring computer program. So, it means that (AI) technology will replace some old computer programs, if human can invent new generation of (AI) software for any functions or activities to satisfy human needs.
As IBM, it argues intelligence. This means that we aim to build systems that enhance and scale human expertise and skills rather than replacing them. We therefore focus on practical applications of (AI) capabilities that assist people in performing well-defined tasks of needs by exploiting and wide range of (AI)-based services. We also use the term " cognitive computing" it is mean a comprehensive net of capabilities based on technology. It comprises the fields of

machine learning, reasoning and decision technologies, language, speech and vision recognition and processing technologies, high performance and high efficient functions for any industries or individual consumers needs. For example, robotics, which are usually very good at doing what which are supposed to in any environment, much have public shopping center, factory etc. places which need simply services from the robot (machine man), such as cleans the floor of our houses to the robot that can work together with humans in production chains, passing through the warehouse, robots can take care of the tasks of an entire warehouse and the companion robots like Nao, Pepper, Aibo and Giraff, who can entertain use, talk to use and help elderly people to stay connected to their friends, relatives and doctors.

Google company is building automatic machine (self-driving cars) and has acquired more than 10 robotics companies. Facebook had opened whole new research facility only on (AI) research. Apply computer has developed Siri. Microsoft computer company has built a similar personalized assistant. Google has Deep mind, a UK company whose long term aim is to build general (AI) and has already great potential to win game to the world champion and IBM is investing a huge amount of resources in applying its Watson cognitive computing system to the medical domains to finance and to personalized education. In Europe, IBM is establishing new centers in Munich and Milan focused in the application of cognitive computer capabilities to the internet of things and healthcare respectively.

For example, automatic machine man (self-driving cars) are all about (AI), which used to be able to see what happens in the street (signals ,lanes, other cars, pedestrians, traffic lights, which need to able predict what other cars and pedestrians will do, and who need to be able to cope with unforeseen situations. Since, most car accidents are due to human fault, it is estimated that the adoption of self-driving cars will save about half of the lives that are usually last in car accidents.

IBM Watson company has to understand spoken language, make

sense of massive amount to text , respond correctly to questions in many categories, as well as assess its own confidence in responding to such questions. In the future, (AI) technology can own question/ answering capabilities that would be very useful, for example, in assisting a doctor when trying to some to the correct diagnosis for a patient and to propose the best therapy .

Intelligent machines can also rely on huge amounts of data to be used to learn how to make better decisions. This data comes from all of us over the years Facebook users have uploaded more than 250 billion pictures and every day who upload about 350 million more. Every second, we submit 40,000 google search queries. So, (AI) technology will be connected through the web from appliances to traffic lights from cars to watches. Other tasks that are very easy for humans are physical and manipulation tasks, such as walking , running, picking up an object to make its shape and location, restricted environment. But (AI) machine man technology still not able to have the general physical and manipulation capabilities even of a 6 year old.

So, it brings this question: Why do (AI) scientists need to concern ethics? Because (AI) technology is complex, information into insight has the potential to reveal long held secrets and help solve some of the world's most difficult problems. (AI) systems can potentially be used to help discover insights to treat disease, predict the whether, and manage the global economy. So, ethic issues is important to and (AI) scientists . If any one new (AI) technology research investigation could success, it will be a secret to and the (AI) scientists can not permit to their loyalty to any competitors to damage the fair (AI) technology products trading market. The country (countries) (AI) technology scientists need to concern ethic issues, who need to keep secrets for their countries economic or/and social benefits. This is moral issues to any countries/country loyalty is whose countries intangible assets. They can not sell (AI) loyalty to any their countries to assist whose economic benefits immorally.

- How can (AI) technology influence to global health care economy development?

According to (AI) lecturer analysis, when combined key clinical health (AI) application can potentially create $150 billion in annual savings for the US healthcare economy by 2026 year. (AI) technology is re-winning modern conception of healthcare delivery. It enables machines to sense, comprehend, act and learn. So which can perform administrative and clinical healthcare functions (Accenture, 2017).

It will help health care service organizations to reduce health care cost, will improve and raise service quality and access. So, (AI) health market size will be predicted growth. (AI) applications in health care include robot-assisted surgery, virtual nursing assistant, administrative workflow assistant, fraud detection, error reduction connected machines, clinical trial participant identifier, preliminary diagnosis, automated image diagnosis and cybersecurity.

What kind of benefits (AI) technology can contribute to healthcare service? (AI) technology can deliver what many health care organizations need, such as financial and operational of labor costs, digital expectations from patient consumers how to use (AI) technology to solve interoperability challenges in any healthcare organizations. Also (AI) technology can be applied to wellness an d lifestyle management, diagnostics, delivers financially but also way of organizational and workflow improvement. So, (AI) technology will be continue to become most prevalent and adoption to healthcare organizations , which must need to enhance structure to be position to take full advantages of new (AI) technological capabilities. (AI) technology can change the nature of work and employment is rapidly changing to make the best use of both humans and (AI) talent in healthcare industry in the future. For example, (AI) technology offers a way to fill in gaps and the rising labor shortage in healthcare. According to Accenture analysis, the physicians shortage is increasing. However, (AI) technology will manufacture healthcare machine men to replace physicians in future one day(2017). Hence, (AI) technology will be invented to raise health care service staffs work efficiency and performance in any hospitals or clinics in the future.

In conclusion, (AI) technology will raise efficiency for any service or manufacturing industries in the future, although, it is possible that it will also rise low skillful workers unemployment numbers. But, the most important influence to human technological innovation will be risen and it will influence human life will be changed to be better, e.g. self drive cars, health care physician machine men, machine man cleaners etc. intelligent machine men will be manufactured to serve for our daily life. Furthermore, (AI) technological products will influence countries trading, some low technological development countries manufacturing businessmen can choose to buy any (AI) products to raise whose productivity and efficiency and reducing cost to achieve economic cost saving result. Also, GDP of trading growth income will increase to the (AI) products sale countries. Hence, it will be beneficial to economic development to both developed and developing countries both in the future as well as (AI) scientists time and money spending will be valued to continue to invest (AI) technology development for human life and economy benefits for long term.

In conclusion (AI) technology will raise macro economy growth and it can create many (AI) jobs , but it also raise the low level technological worker unemployment change. In the future, (AI) technology can be applied to digital technology to attempt to invent any new undiscovered (AI) and digital technology. So, it needs any scientists to continue to research how digital and (AI) technology can be mixed to satisfy human's future undiscovered needs.

III

AI third stage development

Artificial intelligence education development
or the future of defense choice

Nowadays, artificial intelligence (AI) is widely knowledge to be one kind of the dramatic technology. However, it is expected to continue, to have a disruptive impact on human's private and public life, so defense and security will be no exception. But how exactly will these be affected ? How will (AI) defense and security is incremental in nature?

To research why artificial intelligence (AI) has possible to be used to cause autonomous weapons by human. We need to understand these three aspects of relationship. They include cybersecurity and artificial intelligence and machine learning and autonomous weapon systems relationship between of them.

Firstly, we need to know what is the mean of artificial intelligence and cyber defense/offense? It means defense of critical networks: real time, pattern finding, anomaly seeking, it must utilize machine (AI) learning algorithms to efficiently, and instantaneously respond to potential network threats as well as it means human on or out of the loop. On the loop : it means anomaly detection: human notified, IT analysis, response. Out of the loop: it means anomaly detection: (AI) decides best method of response: quarantine, honey pot monitoring, hack-back. Thus, it is possible that (AI) can be used , such as autonomous cyber weapon.

What is artificial intelligence and autonomous weapons? Autonomous weapons mean one kind of weapon that can be selected and engaged a target, without intervention by a human operator. Are these machines artificially intelligent? I believe the answer is not, because present weapons systems are not capable of human level reasoning. But, (AI) algorithms are presently employed to process sensor data, monitor system health, take and respond to vocal commands manage data, navigate. This, future autonomous weapons systems will require stronger (AI) to be secure and operationally and cost effective. Moreover, self-aware autonomous cyber systems are crucial.

What is cybersecurity mean? It means the ability to control access to networked systems and the information they contain. It is acted to prevent , detect, recover, react. It is application objects concern people, process, technology and it's application goals are confidentiality, integrity and popular availability. Thus, what is cyber weapon mean? Walware means viruses, Trojans, zero-days, worms ransomware, spyware etc. Does it require a particular objective? E.g. military paramilitary or intelligence. Does it require physical harm? E.g. functional harm or interruption? Mental harm? Is (AI) a technological weapon that it is an object or tool? What about when it is an weapon agent?

In simplicity, (AI) can be one of scientific weapons platform. When one day, it is invented to be applied to control war planes to fly to any countries to attack enemies or it is invented to be seemed to human to replace soldiers to bring guns or any weapons go to other countries to attack. So, it is possible that future any war defense planes, (AI) technological automatic control weapon can be replaced of human soldiers or war plane pilots to control any war defense planes to go to different enemy countries to attack them easily. It is very horror matter to threaten global human's ourselves life in the future , if (AI) automatic control war defense planes or (AI) automatic control machine soldiers were invented successfully.

Hence , when (AI) can be applied to weapons platforms, it structures that launch weapons, i.e. jets, ships, vehicles. (AI) platform and weapon and software architecture components are be done one (AI) technological weapons systems. Thus, human will encounter any (AI) benefits or risks (threats) causes in the same time as soon as possible. If we can predict when (AI) weapon system will be manufactured or invented successfully. Then, we can reduce (AI) weapon systems risks , if we can threaten any (AI) scientists continue to invent any undiscovered (AI) weapons in any time to avoid the future first time (AI) weapon war occurrence in possible.

The (AI) weapon system risk means autonomy: the ability to problem solve technological war , when (AI) weapon system is manufactured successfully, the power to act, how to damage the (AI) weapon system. The power to chance to stop (AI) weapon system manufacturing processes, ability to create a new goals, how to change the (AI) weapon system inventors' or scientists' minds to avoid to apply (AI) tools to achieve attack goals to change to another positive goal. Due to human can't know a prior what an autonomous (AI) weapon system will do.

Although, human is known what (AI) is , but human is also known when (AI) scientists whose emergent behaviors will do to change to do any negative behaviors from positive behaviors. Whatever (AI) weapon system design we use, there will be cybersecurity, problems arising from computation design/ complexity. Due to any one (AI) scientist can manipulate the system to act against itself, or who can utilize traditional " cyber weapons" against the (AI) weapon system, or who can manipulate the system to lie to humans, but also due to complexity, there is no way to know if it is lying or not or bounded rationality : satisficing.

Finally, the most serious (AI) technological invention risks are human is unknown these aspects of (AI) absolutely: They are not simple automatic systems, learning reasoning, communication of " self-aware" systems. Thus, human will face (AI) technological

invention risks or threats. We need to find any methods to avoid (AI) weapon system is manufactured successfully to avoid (AI) technological war can occur in future anyone day. Consequently, why (AI) scientists do not choose to invent (AI) machine men own human reading, writing, speaking, judgement, analyizing abilities to develop human's future education industry, but they choose to invent (AI) machine men own human weapon to attack enemy ability.
Hence, (AI) scientists have moral responsibilities to avoid to invent (AI) machine men to own human's attack abilities. I shall indicate future what aspects of (AI) machine men can be applied to different education industry aspects as below:

8.1 Online technology and online book technology influences artificial intelligence mind development

Nowadays, online technological invention bring online book technological development. Also, artificial intelligent technological machine men had been invented to link internet to do any jobs, e.g. children can find any data from artificial intelligent machine men when the artificial intelligent machine man had been installed internet and computer function, then children can find any online books to read from the artificial intelligent machine man. Such as Japan artificial intellgent machine men had installed computer and internet function, the Japan family children can find any online books to read from the artificial intelligent machine man at Japan any families' homes conveniently. Hence, it implies that future one day, artificial intelligent machine has possible to be invented to own human's reading and/or writing abilities.
For example,online book publishing is one kind of popular internet technology. For example, Amazon publish is as a business model with many potential advantages, relative to a physical operation. It held out the potential of lower book inventing and distribution costs and reduced overhead. Consumers could find the books, they were looking for more easily and a variety book topic choices could

be offered for sale. It can accept and fulfill orders from almost any domestic location with equal ease. And most purchasers made on its site would be exempt from sales tax. One Amazon strategy hand, it would have to make its returns and redress processes transparent and reliable, and offer other ways for clients to learn, as much about the book possible before buying. Future online book market development trend, such as Amazon, Barnes & Noble etc. online book shops.

Hence, online book store technology can be applied to artificial intelligent technology. Such as artificial intelligent machine men can apply computer technology to learn the abilities of reading and/ or writing any books either on paper or on computer. Hence, it is possible that artificial intelligent machine men will have similar human's writing and/or reading books ability when they own human's mind ability. However, it bring this questions: Can artificial intelligent machine men own human's mind abilities? If they own human's mind abilities, is it mean that they can write and/or read any books? Can artificial intelligent machine men own human's mind abilities to create to write any books? Can artificial intelligent machine men own human's mind abilities to read and make any judgements or decisions more accurate than human's judgements or decisions? To answer these questions? I shall indicate that online book reading and writing technology can be applied to artificial intelligent machine men reading and writing technology. Because they are similiar computer mind technological development. So, I believe that future artificial intelligence machine men can be invented to own similar human's reading and writing's mind abilities in future one day.

I believe artificial intelligence and online technological reading abilities are very similiar. Nowadays, computer can be invented to attempt to read and write any books by human. Why can not artificial intelligent machine men replace computer to read and write any books? Artificial intelligent machine men can replace human to attempt to write or/and read books, due to artificial intelligent machine men had invented to own human mind to do some jobs and their mind had been invented to be similiar to human behavioral abilities to do these behaviors, e.g. cooking, driving, playing games, singing songs, speaking, listening, frighting etc. different human's abilities. So, it seems that artificial intelligent will be possible to be invented to own human's mind abilities to do any writing or reading behaviors or functions.

8.2 Prediction of artificial intelligence reading and writing abilities

development

What is future trend of artificial intelligence reading and writing abilities development? To answer this question, we need to know what benefits of artificial intelligent machine men can attribute to human's needs when they can own any human's mind to read or/ and write any books.

I shall indicate e-books reading and writing example, if artificial intelligent machine men can be invented to own human's mind to write and/or read e-books on computer. Then, it brings this question: Can artificial intelligent machine men assist human to learn to do judgement to solve any challenges?

I believe that when artificial intelligent machine men can be invented to own human mind to write or/and read any books, then they will own human's mind ability to make judgement to solve any challenges more accurately, even their decisions can be more accurate to compare to human's decisions. So, artificial intelligent machine mens' writing and reading ability is the main factor to cause their mind to do any judgement in order to make any decisions more accurately. Consequently, in future one day, artificial intelligent machine mens' writing and reading ability will be invented to similar human's reading and writing abilities as well as their minds can also be invented to similar human's minds as well as their judgement abilities can be invented to similar to human's judgement abilities to make any decisions more accurate.

8.3 The influences when AI is invented to own
human's mind and judgement abilities

Finally, I shall discuss what are the influences when AI is invented to own human's mind and judgement abilities in our future job market. The achievement of artificial intelligent (AI) machine men achievement requirement of owning human's mind and judgement abilities which requires extensive manual labor, and by augmenting the calling process with machine learning, the process where speed and accuracy are needed to close to human's mind and judgement abilities. Expert human race callers now have better information at artificial intelligent machine men at their fingertips faster.

Hence, if the above those requirements are achieved to satisfy artificial intelligent machine men ind and judgement abilities demand to close or exceed humans' mind and judgement abilities. Then, I believe that future human's some simple jobs must be replaced by (AI) machine men. Even, human's some professonal jobs, e.g. lawyer, accountant, administator, typing etc. professional skillful jobs, which will be either replaced or will be assisted by (AI) machine men. For example, (AI) machine men learn how to type english or other language words to do typing job ; they can learn how to apply accounting knowledge to record any firm's income and expenditure record of accounting job; they can also learn how to assist architects to design any architectural building drawing plans to do architect jobs; they can learn how to analyze any court evidences to judge any criminal or civil cases and assist lawyers to give legal advices to achieve more reasonable judgement for any legal cases; they can also learn how to assist firm's managers or administrators to manage any organization teams efficiently.

Consequently, when (AI) machine men can be invented to achieve to exceed human's mind and judgement abilities level. Then, I believe that they can do instead of human' simple jobs, which can do even human's more difficult and more judgement requirement of professional skillful jobs. So, (AI) machine men must need to achieve to do any jobs, they are same, even exceed to human professionals' abilities. Then, it will cause a lot of human's jobs to be disappeared or some human's jobs will be replaced by owning judgement and mind abilities of (AI) machine men to do.

Hence, future many human's jobs will be replaced by technological labors. Employers choose to buy (AI) machine men to replace human labors. The reasons include (AI) machine men have none unhappy, angry emotin to influence their low efficiencies and low productivities. Their judgement and mind abilities can exceed human's abilities or do any jobs to compare better performance to human's abilities. Consequently, different occupation labors need to prepare to learn how to co-operate with (AI) machine men to let future employers feel (AI) machine men will be human's assistant to assist human to do jobs efficiently when human and (AI) machine men work together. It aims to avoid future employers feel (AI) machine men's judgement and mind abilities can exceed any low knowledgeable and skilful occupation labors, even high knowledge and skilful occupation labors. It means that (AI) machine men are only labors' assistant if (AI) machine mens' judgement and mind abilities are below under to human labors' judgement and mind abilities.

Consequently, to avoid (AI) machine men can replace human to do any simple or complex jobs to cause any future any occupation labors' competitiors. I recommend that it is right time labors ought prepare to learn different skills. So, every individual labor does not only concentrate on one kind of skill. Because supposing one kind of the occupation labor's job duties are replaced by (AI) machine men. If the employee had owned more than one kind of occupation skill. Then, I believe that who can avoid the unemployment threat more easier than the employee only owned one kind of occupation skill, when (AI) machine men had invented to own human's mind and judgement abilities in future one day. The most important, I believe that (AI) invention will be applied to be teach how to learn human's skills and mind ability. Such as education industy, teacher won't be replaced by (AI), otherwise, (AI) will be teacher's assistant to help them to do education data gather or teaching jobs. So, teachers won't be replaced by (AI), otherwise, teachers will depend on (AI) data gather or teaching job to give them opinions how to solve student's teaching challenges as well as teachers can concentrate on researching education jobs for schools' benefits if (AI) technology can be invented to on human's mind and judgement and reading and writing abilities in the future.

8.4
Artificial intelligence and the future of defense or teaching choice

Nowadays, artificial intelligence (AI) is widely knowledge to be one kind of the dramatic technology. However, it is expected to continue, to have a disruptive impact on human's private and public life, so defense and security will be no exception. But how exactly will these be affected ? How will (AI) defense and security is incremental in nature? If (AI) technological machine men are applied to teach students in education aspect, is it better to my next generation learning develpment more than they are applied to war attack aspect.

To research why artificial intelligence (AI) has possible to be used to cause autonomous weapons by human. We need to understand these three aspects of relationship. They include cybersecurity and artificial intelligence and machine learning and autonomous weapon systems relationship between of them. Basic on (AI) machine can be invented to learn any new knowledge, so if (AI) machine men are taught how to attack enemy, which will be such as human soldier function. But, if (AI) machine men are taught how to learn university knowledge to teach students. Then, they will be such as human lecturer function. So, when (AI) is invented to own human mind and judgement and learning abilities, then they will be either human's enemy or human's assistant, such as university lecturer's assistant.

Firstly, we need to know what is the mean of artificial intelligence and cyber defense/offense? It means defense of critical networks: real time, pattern finding, anomaly seeking, it must utilize machine (AI) learning algorithms to efficiently, and instantaneously respond to potential network threats as well as it means human on or out of the loop. On the loop : it means anomaly detection: human notified, IT analysis, response. Out of the loop: it means anomaly detection: (AI) decides best method of response: quarantine, honey pot monitoring, hack-back. Thus, it is possible that (AI) can be used , such as autonomous cyber weapon. If (AI) is applied to make the decision best method of response to learning aspect, such as univeristy different subject knowledge. Then, it will be one good technological educational tool to teach university students.

In simplicity, (AI) can be one of scientific weapons platform or one of university teaching tool. When one day, it is invented to be applied to control war planes to fly to any countries to attack enemies or it is invented to be seemed to human to replace soldiers to bring guns or any weapons go to other countries to attack. So, it is possible that future any war defense planes, (AI) technological automatic control weapon can be replaced of human soldiers or war plane pilots to control any war defense planes to go to different enemy countries to attack them easily. It is very horror matter to threaten global human's ourselves life in the future , if (AI) automatic control war defense planes or (AI) automatic control machine soldiers were invented successfully. Otherwise, when one day, (AI) machine lecturer is invented to be applied to learn university different subjects knowledge to replace lecturers to copy lecturer's every prepared lecturer course to speak to let students to listen when they are sitting in university halls as well as the (AI) machine lecturer can make analysis and judgement response to answer every student's enquire immedicately after it had speaking all courses to students to listen in lecturer hall every time. Then, it can let human lecturer does any education job duty, e.g. research education work. So, (AI) machine lecturer will be future human lecturer's assistant in future one day.

Finally, the most serious (AI) technological invention risks are human is unknown these aspects of (AI) absolutely: They are not simple automatic systems, learning reasoning, communication of " self-aware" systems. Thus, human will face (AI) technological invention risks or threats if human invent (AI) machine man to learn how to attack enemy. Otherwise, if human invent (AI) machine man to learn how to teach univerity student. I believe that my future university students can raise learn ability and writing ability and reading ability from (AI) machine lecturer teaching more than human lecturer teaching.

8.5
Online technology and online book technology influences artificial intelligence mind development

Nowadays, online technological invention bring online book technological development. Also, artificial intelligent technological machine men had been invented to link internet to do any jobs, e.g. children can find any data from artificial intelligent machine men when the artificial intelligent machine man had been installed internet and computer function, then children can find any online books to read from the artificial intelligent machine man. Such as Japan artificial intellgent machine men had installed computer and internet function, the Japan family children can find any online books to read from the artificial intelligent machine man at Japan any families' homes conveniently. Hence, it implies that future one day, artificial intelligent machine has possible to be invented to own human's reading and/or writing abilities.

For example,online book publishing is one kind of popular internet technology. For example, Amazon publish is as a business model with many potential advantages, relative to a physical operation. It held out the potential of lower book inventing and distribution costs and reduced overhead. Consumers could find the books, they were looking for more easily and a variety book topic choices could be offered for sale. It can accept and fulfill orders from almost any domestic location with equal ease. And most purchasers made on its site would be exempt from sales tax. One Amazon strategy hand, it would have to make its returns and redress processes transparent and reliable, and offer other ways for clients to learn, as much about the book possible before buying. Future online book market development trend, such as Amazon, Barnes & Noble etc. online book shops.

Hence, online book store technology can be applied to artificial intelligent technology. Such as artificial intelligent machine men can apply computer technology to learn the abilities of reading and/or writing any books either on paper or on computer. Hence, it is possible that artificial intelligent machine men will have similar human's writing and/or reading books ability when they own human's mind ability. However, it bring this questions: Can artificial intelligent machine men own human's mind abilities? If they own human's mind abilities, is it mean that they can write and/or read any books? Can artificial intelligent machine men own human's mind abilities to create to write any books? Can artificial intelligent machine men own human's mind abilities to read and make any judgements or decisions more accurate than human's judgements or decisions? To answer these questions? I shall indicate that online book reading and writing technology can be applied to artificial intelligent machine men reading and writing technology. Because they are similiar computer mind technological development. So, I believe that future artificial intelligence machine men can be invented to own similar human's reading and writing's mind abilities in future one day.

I believe artificial intelligence and online technological reading abilities are very similiar. Nowadays, computer can be invented to attempt to read and write any books by human. Why can not artificial intelligent machine men replace computer to read and write any books? Artificial intelligent machine men can replace human to attempt to write or/and read books, due to artificial intelligent machine men had invented to own human mind to do some jobs and their mind had been invented to be similiar to human behavioral abilities to do these behaviors, e.g. cooking, driving, playing games, singing songs, speaking, listening, frighting etc. different human's abilities. So, it seems that artificial intelligent will be possible to be invented to own human's mind abilities to do any writing or reading behaviors or functions.

8.6 Prediction of artificial intelligence reading and writing abilities

development

What is future trend of artificial intelligence reading and writing abilities development? To answer this question, we need to know what benefits of artificial intelligent machine men can attribute to human's needs when they can own any human's mind to read or/ and write any books.

I shall indicate e-books reading and writing example, if artificial intelligent machine men can be invented to own human's mind to write and/or read e-books on computer. Then, it brings this question: Can artificial intelligent machine men assist human to learn to do judgement to solve any challenges?

I believe that when artificial intelligent machine men can be invented to own human mind to write or/and read any books, then they will own human's mind ability to make judgement to solve any challenges more accurately, even their decisions can be more accurate to compare to human's decisions. So, artificial intelligent machine mens' writing and reading ability is the main factor to cause their mind to do any judgement in order to make any decisions more accurately. Consequently, in future one day, artificial intelligent machine mens' writing and reading ability will be invented to similar human's reading and writing abilities as well as their minds can also be invented to similar human's minds as well as their judgement abilities can be invented to similar to human's judgement abilities to make any decisions more accurate.

8.7 The influences when AI is invented to own human's mind and judgement abilities

Finally, I shall discuss what are the influences when AI is invented to own human's mind and judgement abilities in our future job market. The achievement of artificial intelligent (AI) machine men achievement requirement of owning human's mind and judgement abilities which requires extensive manual labor, and by augmenting the calling process with machine learning, the process where speed and accuracy are needed to close to human's mind and judgement abilities. Expert human race callers now have better information at artificial intelligent machine men at their fingertips faster.

Hence, if the above those requirements are achieved to satisfy artificial intelligent machine men ind and judgement abilities demand to close or exceed humans' mind and judgement abilities. Then, I believe that future human's some simple jobs must be replaced by (AI) machine men. Even, human's some professonal jobs, e.g. lawyer, accountant, administator, typing etc. professional skillful jobs, which will be either replaced or will be assisted by (AI) machine men. For example, (AI) machine men learn how to type english or other language words to do typing job ; they can learn how to apply accounting knowledge to record any firm's income and expenditure record of accounting job; they can also learn how to assist architects to design any architectural building drawing plans to do architect jobs; they can learn how to analyze any court evidences to judge any criminal or civil cases and assist lawyers to give legal advices to achieve more reasonable judgement for any legal cases; they can also learn how to assist firm's managers or administrators to manage any organization teams efficiently.

Consequently, when (AI) machine men can be invented to achieve to exceed human's mind and judgement abilities level. Then, I believe that they can do instead of human' simple jobs, which can do even human's more difficult and more judgement requirement of professional skillful jobs. So, (AI) machine men must need to achieve to do any jobs, they are same, even exceed to human professionals' abilities. Then, it will cause a lot of human's jobs to be disappeared or some human's jobs will be replaced by owning judgement and mind abilities of (AI) machine men to do.

Hence, future many human's jobs will be replaced by technological labors. Employers choose to buy (AI) machine men to replace human labors. The reasons include (AI) machine men have none unhappy, angry emotin to influence their low efficiencies and low productivities. Their judgement and mind abilities can exceed human's abilities or do any jobs to compare better performance to human's abilities. Consequently, different occupation labors need to prepare to learn how to co-operate with (AI) machine men to let future employers feel (AI) machine men will be human's assistant to assist human to do jobs efficiently when human and (AI) machine men work together. It aims to avoid future employers feel (AI) machine men's judgement and mind abilities can exceed any low knowledgeable and skilful occupation labors, even high knowledge and skilful occupation labors. It means that (AI) machine men are only labors' assistant if (AI) machine mens' judgement and mind abilities are below under to human labors' judgement and mind abilities.

Consequently, to avoid (AI) machine men can replace human to do any simple or complex jobs to cause any future any occupation labors' competitiors. I recommend that it is right time labors ought prepare to learn different skills. So, every individual labor does not only concentrate on one kind of skill. Because supposing one kind of the occupation labor's job duties are replaced by (AI) machine men. If the employee had owned more than one kind of occupation skill. Then, I believe that who can avoid the unemployment threat more easier than the employee only owned one kind of occupation skill, when (AI) machine men had invented to own human's mind and judgement abilities in future one day.

8.8
Why does AI machine lecturer can raise education quality?

When (AI) machine men can own human reading and writing and judgement and analytical abilities, then they can replace university lecturers to teach students to raise students' learning abilities absolutely. Then, it bring this question: Why does I machine lecturer can raise education quality? Why do universities prefer to apply (AI) machine lecturer to teach teachers more than human lectuer in university lecturer hall learning environment? Will (AI) university lecturers replace human lecturers to teach students to learn at university lecturing halls popularly? Can (AI) university lecturers replace university human lecturers to teach students more easily and it can let students feel more easily to learn when they are listening what (AI) university lecturers are teaching to them every time university lecture.

In university today, nearly all students need to attend university lecturing hall to listen their lecturer's teaching in every time course. However, many students do not feel interesting to attend university halls to listen human lecturer's teaching. The reasons include, they are busy, so no time to attend lecturer's hall to listen lecturer's teaching; or they feel bore to listen their lecturer's teaching; they feel difficulty to learn; they have confidence to exam and do their assignments, so they feel that they do not need to go to lecturing halls to listen their human lecturer's teaching. However, if one day, (AI) machine lecturers are invented to teach university students to learn and solve their learning difficulties. Can it raise student individual learning interest, due to (AI) machine lecturers' education quality is better than human lecturers' education quality?

What will influence to university students if (AI) machine lecturer can invented to replace human lecture? The influences will include such as below:

First reason: the only way is going to be useful to university lecturers are if all (AI) machine lecturers are well-informed and fully supported to assist human lectuers to teach whose students to let them to listen whose teaching absolutely. So, human lecturers can concentrate on doing any education research and data gathering jobs to prepare for (AI) machine lecturers to help them to explain human lecturers' every time prepared course contents more efficiently. So, (AI) machine lectuers can help human lecturers to share whose teaching time in lecturing halls. Human lecturers' can spend whose hall lecturing time to do whose educational research or other educational gathering jobs absolutely.

The second reason, the human lecturer (Human capital) has ability and efficiency of concentrating on education data gatehering research jobs to prepare to write whose books. When (AI) machine lecturer replace the human lecturer to spend time to attend lecturing hall to teach students. Fo long term, the human lecturer can raise education productivity growth and education quality raising, due to who only concentrate on searching or gathering data to prepare to write whose books to raise their education level.

In macro and micro economic view, the well (AI) machine lecturer educated labor (human capital) is often replaced to human lecturer as one of the critical factors to influence rapid education productivities and educational quality growth to the Asia developing countries' any regions or cities. Because any of these Asia developing countries, such as China, Korea, Philippines etc. countries which need have well educated and knowledgeable lecturer labors to raise any universities' educational productivities and educational qualities growth. So (AI) assistant lecturer factors which ought have close relationship to cause the good or bad future student learning effectiveness and education or learning qualities raising in these any one of Asia developing countries.

The third reason, for the big population of student growth number example, China's student growth rate is larger than school growth rate. If China expect every students have enough chance to study in schools, but university human lectuer numbers are not enough to supply to universities to teach their students. I believe that (AI) machine lecturer is only one kind of teaching method to solve these big population countries' university lectuer number shortage challenge.

In conclusion, in long term, (AI) machine lecturers can solve university human lecturer shortage challenge as well as they can attract many students to attend lecturing halls and human lecturers can raise education quality when they can concentrate on searching or gathering data to prepare their education career, when (AI) machine lectuers replace them to spend time to attend univesity halls to teach students in every university lecturing time.

8.9

Future AI machine education market

I believe that when AI (artificial intelligent machine men) which can invented to own to similar to human mind, learning, language, analytical, judgement abilites. Then, which can be applied to any education market service industy. (AI) potential education market service industy includes such as below:

- (AI) university lecture assistant

Future (AI) machine men can assist univerity lectuers to attend university halls to attempt to teach university students for different subjects, e.g. english, math, economic, math, engineering, art, architect etc. different subjects. It depends on the human lectuter who prepares to spend time to teach the (AI) machine lecturer to remember whose teaching subject. For example, the economic lecturer spend one year time to teach the (AI) machine lecturer to learn all economic knowledge. Then, the (AI) machine lecturer can use its machine brain to remember all the human lecturer's economic concepts and prepared teaching economic contents within the one year. Hence, after one year the (AI) machine lecturer can remember all the human lecturer's economic concepts and economic theories and economic contents to prepare to attend university lecturing halls to teach all first year undergraduated first year economic students confidently. It means that the human lecturer's job duties will change to teach (AI) machine lecturer to learn whose economic knowledge to prepare to let the (AI) machine lecturer to replace whom to teach whose university students; so the human lectuer can spend more time to do other research job for whose university education development. Hence, the (AI) machine lecturer can share the human lecturer teaching job as well as the human lectuer can concentrate on spending time to do whose research jobs for whose university education development. This is one both win strategy to university and the lecturer if (AI) machine lecturer is invented to assist future university lecturer's teaching jobs

- (AI) secondary and primary teacher assistant

In the future (AI) technolgical development, instead of (AI) machine men can be applied to university education aspect. Future (AI) machine men can also be applied to secondary and primary teaching aspect. For example, primary and secondary schools do not need attend classroom to teach students. (AI) machine teachers can replace them to attempt to do teaching job. They only need to spend one year time to prepare to teach (AI) machine teacher to learn how to apply their teaching skill concern their subjects who need to teach to their students, e.g. english language writing and reading and spelling skill, sing song skill, drawing picture skill, calculation skill etc. different studying skill. Then, the (AI) primary or secondary machine teacher can apply the primary or sendary human teacher skills to attempt yo teach whose students. Hence, the primary and secondary human teacher whose duties will change to learn how to teach whose teaching skills to let the (AI) primary or seondary machine lecturer to remember how to apply human skills to teach whose students for different subjects, such as, english writing and reading and spelling language skills, singing songs language skills, math calculation skills etc. Hence, future primary or secondary school teachers who responsibilities will change to learn how to teach (AI) machine teacher teaching skills to prepare to replace them to teach their students in classroom.

● (AI) scientific research assistant

Future (AI) machine men can be applied to science research aspect, instead of school education job. For example, (AI) machine men can be any scientist's assistant, e.g. space scientist, earth or ocean scientist, human or animal behavioral psychological scientist, climate scientist, chemical scientist, drug scientist etc. How can (AI) machine men can be any kind of scientist to assist scientists to do research jobs ? I shall indiate such as below:

For space science example, the (AI) machine space scientist can assist human space scientist to gather space data to assist space scientist to research any undiscovered material to cause our earth, even space. Hence, the space scientist only need to teach the (AI) machine scientist to learn how to help them to apply space technological tools to gather data and then enter all data to computer to record, even the (AI) machine scientist can store all space data discovered record to their machine brain every day. Hence, the human space scientist does not need to spend much time to do gathering data job. The (AI) machine space scientist can help whom to do these space data gathering job, then the human space scientist can concentrate on spendin time to do space research job in whose space science laboratory every day.

For earth science example, the (AI) machine earth scientist can help the earth scientist to go to anywhere to gather earth or ocean natural activity data in our earth every day. Then, the earth scientist only need sit in whose earth laboratory to wait the (AI) machine earth scientist to come to whose laboratory to give whose gathering every day earth or ocean natural activity data to do future research job. Hence, the earth scientist does not need to leave whose laboratory to do any data gathing jobs concern earth or ocean natural activities. The (AI) machine earth or ocean scientist had helped him/her to go to our earth or ocean anywhere to do any earth or ocean activities data gathering jobs every day. Hence, the earth or ocean scientist can concentrate on spending whose time to do any research jobs in laboratory. It means that the (AI) machine earch or ocean scientist had replaced whom to do all outdoor original gathering data jobs.

For these human or animal behavioral psychological scientist, climate scientist, chemical or drug scientist, scientist all examples, the (AI) machine human or animal behavioral psychological scientist can help them to do any data gathering job, e.g. the (AI) machine scientist can learn how to help human or animal behavioral psychological scientist to contact human or animal to observe their daily activities and record all their activities data to transfer all these daily activites data to let the human or animal behavioral psychological scientist to do psychological researching analysis only. The (AI) machine climate scientist can help the human climate scientist to arrive anywhere to observe climate changes and record climate changes daily. Then, the human climate scientist only need to wait the (AI) machine climate scientist's gathering climate change data record from whose machine brain to do climate changing predict research job in climate laboratory every day. The (AI) chemical or drug machine scientist can help the drug or chemical scientist to gather data of new drug or chemical from internet channel every day. So, the human chemical or drug scientist only need to do researching job after the (AI) machine scientist transfers all daily chemical or drug information to let them to know from internet channel. It means that the chemical or drug human scientist does not need to spend much time to gather drug or chemical new data development trend from internet. The (AI) machine chemical or drug scientist had helped them to do data gathering job every day.

Consequently, future (AI) machine men can do education and research aspects of jobs duties and their role are only human scientists or primary or secondary teachers or university lecturers whose assistants either to share scientist's data gathering job or share teachers or lecturers' teaching job.

Artificia intelligent future education market development

9.1 Artificial intelligent robots invention

negative impacts

Indeed, artificial intelligence, computing can learn to something that effectively reasons, thinks, if (AI) learns more powerful and valuable complement to human capabilities; improving medical diaguoses, weather prediction, supply-chain management, transportation and even personal choices about where to go on vacation or what to buy, how to learn teaching any subjects knowledge to assist school teachers to teach students, e.g. accounting, law, architecture, language etc. subjects knowledge It can bring benefits to human it (AI) robots can learn teacher's any subjects to teach students or skillful knowledge, e.g. driving, taking care old people at home, manufacturing vehicles in factories. Otherwise, if (AI) robots learn how to be applied to be weapons to attack enemy. It will bring harm to human's life safety. As a result, technology can assist human's development when it can be applied to do meaning jobs, but it can also damage human's development when it can be applied to do harmful human's behaviour, e.g. attacking enemy to cause robot technological war. Consequently, the negactive impact will be depending on how humans teach (AI) robots to learn when some humans intends to teach (AI) robots to attack enemy. Thus, scientists need to know to educate (AI) robots to learn positive knowledge to attribute to us to raise our standard of living, if it is a tool in the service of humans, making our lives better. Otherwise, who ought not educate (AI) robots to learn negative knowledge to attribute to influence our's life safety when they are applied to attack enemies.

However, some scientists believe (AI) robots will have negative impact to influence our economy and society. Such as (AI) robots will destroy most jobs, it will make humans foolish, due to humans depend on (AI) robots to assist us to do any jobs; it will destroy people's privacy and it will enable bias and abuse and it will eventally exterminate humanity. Hence, when time comes to develop computers really think, intelligence is the same things as consciousness, the brain is a computer. Then, scientists must have responsibilities to concern how to teach (AI) robots to learn useful or attributable human's knowledge, not harmful human's knowledge to avoid future invented (AI) robots are taught to learn how to attack ourselves.

9.2 Future AI tutoring system potential development market

Humans can learn (AI) robots how to educate our next generation after it has be taught any knowledge from human. Thus, it brings this question: How can human teach new knowledge to (AI) robots to learn successfully? I shall indicate some scientists' evidences to explain that it is possible (AI) robots had abilities to learn human's mind, judgement and analytical abilities, reading, writing, speaking abilities in future one day absolutely. I shall indicate what is intelligent tutoring systems (ITS) example as below:

When are computer-based tutors which act as a supplement to human teachers. The major advantage of an (ITS) is, it can provide personalized instructions to students according to their cognitive abilities. In this scenario, an intelligent tutoring system can be quite relevant to solve unavailability of skilled teachers challenges. Such as India has many students who demend teachers to learn to them, but India is facing skilled teachers shortage challenge. (ITS) are computer-based tutors, which act as a supplement to human teachers. An intelligent tutoring system is educational software containing an artificial intelligence component. The software tracks students work, tailoring feedback and hints along the way. By contenting information on a particular student's performance, the software can make interences about strengths and weaknesses, and can suggest addition work.

The ITS's one of main advantages is individualized instruction delivery to the group of same learning ability of student in evey classroom, which means the system ill adapt itself to different categories of students. A real classroom is usually heterogeneous where these are different kinds of students, from slow learners to fast learners. It is not possible to provide attention to them individually. Thus, the teaching may not b beneficial to all students. An ITS can eliminate this problem, because in this virtual learning environment, the tutor and the student has a one-to-one relationship. When the school applies TIS intelligent tutoring systems to teach its students, the students can learn in whose own method. Another advantage is their using this system teaching can be accomplished with there is trained teachers. Hence, the functionalities of this intelligent tutor system can be divided into three major tasks: (1) organization of the domain knowledge, (2) keeping track of the knowledge and peformance of the learner, (3) planning the teaching strategies on the basis of the learner's knowledge state. Hence, to carry out these tasks, the ITS have different modules and interfaces for communication (Wenger, 1987; Freedman, 2000; Chou et. at, 2003).

The criteria of robots in education include domain of the learning activity, location of the activity, the role of the robot, types of robots and types of robotic behavior. Scientists indicate that robots are primarily used to provide language, science or technology education and that a robot can take in the role of tutor, tool or peer in the learning activity. However, robotics can include these kinds, such as social robotics, pedagogy, human robot interaction and educational robotics.

Early, industrial robots are invented to apply to manufacturing industry only in 2008 beginning. robots are slowly beginning a process to everyday, lives both at home and at school (IFR, 2008). Nowadays, the most popular countries accept to apply robots to replace human's some jobs, include Japan, Korea, USA, Australia, Germany, Holland.

What is the domain of the learning activity to robots? To develop robots to education industry, the first criterion of the two main categories are robotics and computer educaion (the awareness of technology that could be referred as technical education) and non-technical education (science and language). The technical education means giving students the knowledge of robots and technology. For example, introducing computer science and programming and to familiarize undergraduate students with technology and schol students were gradually exposed to technical subjects using robots. A lesson plan usually involves first an initial introduction to programming the robot (introduction phase) and then the students apply their knowledge practically by making their robots work (intensive phase. The second observed domain in the area of robots in education are non-robots in education are non-technical subjects (such as the sciences), where schools witness the employment of robots as an intermediate tool to impact some form of education to students in classrooms, such as mathematics. The third common domain is the use of robots to teach a second language. For example, English was taught to Asia countries children by robots in by researchers from the robotics laboratory. For another example, the implication of using robots to teach a second language have been well documented by computer science researchers in Taiwan, where it is stated that children are not as hesitant to speak to robots in a foreign language as they are talking to a human instructor. So, it seems, future robots can be a language teachers to the learning foreign language students. However, the language robots require having accurage speech recognition ability and how to learn in acknowledging the use of robots for language instruction ability in future robots language education market.

9.3 How can potential social teaching robots
assist to teacher in school?

What benefits do (AI) robots provide to education industry? Artificial intelligence can indicate how to improve courses, when teachers may no always be aware of gaps in their lecturers and educational materials that can let students confused about certain concepts. Different students have different learning styles, abilities, interests and needs. For classroom situation example, on teacher in a classroom of 20 to 30 students will rarely be able to cater to each of those needs. Homework and classes could be customized based on a student profile, interests can be cultivated based enhanced by exposing students to different course and content.

Artificial intelligence can offer a way to solve that problem. For online or non-online course providers cases, will have already benefits if these apply (AI) robots to educate. When a large number of students are found to submit the wrong answer to a homework

assignment, the system alerts the teacher and goves future students a customized message that offers hints to the correct answer.
This type of system helps to fill in the gaps in explanation that can occue in online or non-online courses, and helps to ensure that all students are building the same conceptual foundation. Rather than waiting to either hear back from the professor in lecturng hall or classroom from classroom education method or receive back professor's feedback message from the internet online education channel. So, classroom students or online learning students can get immediate feeback that helps them to understand a concept and remember how to do it correctly the next time around if (AI) robots can be applied to assist university professor or primary/ secondary teachers to teach whose students either in classroom teaching learning environment or online teaching learning environment. Thus, (AI) educational and social robots can be attempted to accepted to teach primary school, high school or university students.
How to apply (AI) education robots to teach students in which different educational functional aspects?
Firstly, on grading system hand, artificial intelligence can automate basic activities in education, like grading. In, college, grading homework and tests for large lecture courses can be for much grading activites to secondary teachers or university lecturers to spend much time to do grading jobs for every student. Even in lower grades, teachers often find that grading takes up a significant amount of time, time that could be used to interact with students, prepare for class, or work on professional development. When, (AI) may not every be able to truly replace human grading, it's getting pretty close. it's now possible for teachers to automate grading for nearly all kinds of multiple choice and fill-in-the blank testing and automated grading of student writing may not be far behind. Today, essay grading software will improve over the coming years, allowing teachers to focus more on in class activities and student interface then grading.
Secondly, on (AI) tutor's support hand, students could get additional

support from (AI) tutors. When, there are obviously things that human tutors can offer that machines can't. The future could see more students being tutored by tutors that only exist in zero and ones. Some tutoring programs based on artificial intelligence already exist and can help students through basic, mathematics, writing, accounting, law and other subjects. (AI) tutors can teach students fundamentals for helping students learn high-order thinking and creativity, something that real-world teachers are still required to facilitate. It should not rule out the possibility of (AI) tutors being able to these things in the future. With the rapid technology advancement that has marked the past few decades, advanced (AI) tutoring systems will be popular to be applied to teach students.

Thirdly, on (AI) driven program educators helpful feedback hand, (AI) can't only help teachers and students to craft courses, that are customized to their needs, but it can also provide feedback to both about the success of the course as a whole. Some schools, especially those with online offerings are using (AI) systems to monitor student progress and to alert professors when there might be an issue with student performance. These kinds of (AI) systems allow students to get the support they need and for professors to find areas where they can improve instruction for students who may be given feedback how to learn subject matter. (AI) programs at thes schools aren't just offering advice on individual courses. However, some are working to develop (AI) systems that can help students to choose majors based on areas where they succeed.

Fourthly, on (AI) changing the role of teachers hand, there will always be a role for teachers in education, but what role is and that it entails may change, due to new technology in the form of intelligent computing systems. As (AI) can take over tasks like grade, can help students improve learning, any may even be a substitute for real world tutoring. (AI) systems could be programmed to provide expertise, serving for students to ask questions, and find information or could even potentially replace the teachers for basic course materials. In most cases, however, (AI) will shift the role

of the teacher to that of facilitator. Teachers will supplement (AI) lessons, assist students who and provide human interaction and hands-on experiences for students. Hence, (AI) technology has already changed in the classroom teaching method, specially in schools that are online lessons.
Fifthly, on (AI) trial-and-error learning hand, trial and error is a critical pasrt of learning, but for many students, the idea of failing, or even not knowing the answer. An intelligent computer system, designed to help students to learn, to deal with trial and error. Artificial intelligence could offer studrnts a way to experiment and learn in a relatively judgement, free environment especially when (AI) tutors can offer solutions for improvement. In fact, (AI) is the perfect format for supporting this kind of learning as (AI) systems themselves often learn by a trial and error method.
Sixthly, on changing student individual learning hand, (AI) has the potential to change where students learns, who teaches them, and how they acquire base skills, using (AI) systems, software and support. Students can learn from anywhere in the world at any time and with these kinds of programs taking the place of certain types of classroom instruction. So, (AI) robots may replace teachers in some instances (for better or worse). Educatonal programs is powered by (AI). They are already helping students to learn basic skills, but as these programs grow. Consequently, as (AI) technology can give above advantages to educational organizations. So, in the future, it seems that (AI) can be popular to be used in online or non-online classrooms to assist teacher individual job to raise educational quality.

9.4 Psychological (AI) eduational
social robots and students relationship

Whether can (AI) educational robots bild good social relationsip to students? Some scientists had attempted to do experiments to prove whether (AI) educational robots can do good or bad social relationshp to students.
For example, Knox, W.B. el. (2012) had ever attempted to do two

experiments to research whether (AI) robot educational robot can build better or worse social relationship to compare human robot. Their two experiments aim to ask how differing conditions affect a human teacher's feedback frequency and the computational agent's learned performance. The first experiment considers the impact of a self-perceived teaching role in contrast to believing one is critiquing record. The second considers whether a human trainer will give more frequent feedback if the agent acts less (i.e. choosing actions believed to be worse). When the trainer's recent feedback frequency decreases. From the results of those experiments, they draw three main conclusions that inform the design of agents. More broadly, these two studies indicate as early examples of a nascent technique of using agents as highly specifiable social entities in experiments on human behavior. Thus, it implies (AI) educational robots have ability to learn human teacher to teach students in good social relationship learning environment with students. Even, (AI) educational robots can build better learning relationship to compare human teachers between students, it seems (AI) robots have attractive ability t raise student individual learning interest, after which can applied to assist teachers to teach whose students in classrooms or lecture halls or online classroom channels.

Other scientist had ever attempted to do experiments about " reinforcement learning" (RL) to research how result of interactive supervisory input between human teacher and both robot and software agents relationship. M. Mataric (1997) attempted to do one experiment concerns that reinforcement learning is designed for interactive supervisory input from a human teacher, several works in both robot and software agents have adapted it for human input by letting a human trainer control the reward signal. He aimed to examine the assumption, namely that the human-given reward is compatible with the traditionanl RL reward signal. He described an experimental platform with a simulated RL robot and present an analysis of real time human teaching behavior found in a study in which untrained subjects taught the robot to perform a new task. For the experiment, who reported three main observations on how people administer feedback when teaching a robot a task through reinforcement learning : (a) they use the reward channl not only for feedback, but also for future directed guideance, (b) they have a positive bias to their feedback, possibly using the signal as a motivational channel, and (c) they change their behavior as they develop a mental model of the robotic learner.

Thus, whose experiment concluded that machine learning shall play a significant role in the development of robotic assistants that operate in human learning environment. (e.g. homes, schools, hospitals, offices). Considering the difficulty of hard-coding all information needd for the robot to play a long term role in az dynamic world, human users will need to be able to easily teach such robots. However, various works have addressed some of hard problems robots face when learning in the real-word.

In robots learning process, what difficulties which will face, the scientist indicated this question concerns how to influence robot's learning abilities: Has RL certain desirable qualifties, such as the learning abilities possibility to explore and learn from unsupervised experience? Many also queation RL as a variable technique for learning in complex real-world environments because of practical problems, such as long training time requirements, non-scaling state representations, sparse rewards (resulting in slow utility propagation) and safe exploration strategies. As a result, reinforcement learning has been utilized for teaching robots and game characters, incorporating real-time human feedback by having a person supply reward and/or punishment as an additional input to the reward function.

Consequently, the scientist discovered human's teaching method is the most important factor to influence the robot's learning abilities amonf all other environment factors. He also assumed and argued that reinforcement based learning approaches should be reformulated to move effectively incorporate a human teacher. To do this properly, the educational robot must understand the human teacher's contribution; how does the human teach? and what does the educational robot try to communicate from a robot learner?

The scientist suggested human trainers ought use these methods to educate robot learners to learn more easily. His main findings indicates reward factor can influence the human robot teachers' motives to teach robot learners to learn, due to more reward can encourage the human robot trainers to teach robots to learn their knowledge and skill. He also found that robot users read the behavior of the robot machine learners and adjust the human robot trainer whose training strategies as whose mental model of the different robot human trainer education method changes. Viweing the human input as a traditional RL reward signal does not take advantage of the fact that a teacher adjusts hose training behavior to best suit the robot educational learner.

In addition to the related RL works mentioned above. Every human robot trainer needs to consider the topic of human input for machine learning systems. Personalization agents and adaptive user interfaces are examples of software that learns by observing human behaviors modeling humann preferences or activities. It empathizes how human teaches the robot learner through interaction, various works address trainable software and robotic agents, exploring explicit human input: learning classification tasks and navigation tasks via natural language, robots that learn by demonstration or/and software agents that learn or training. It seems how to design the robot machine learning software technology will also influence the robot's learning abilities. Thus, human trainer's software learning machine and how to give reward to encourage the human trainer's teaching behavior to let the robot to learn, these factors will influence the robot's learning ability to be applied to education industry to assist human teacher's teaching job successfully. Because how much knowledge and skill, the robot can learn from the human trainer, how much educational knowledge that the robot can own to prepare to teach any students to learn easily. Hence, the human trainer's knowledge and skill will quality to satisfy its primary, secondary and university students learning need.

9.5 Can robots be teachable agents to students really?

In the future, I believe robots have potential to contribute to education by acting as subordinate learners for students teaching. The benefits that the act of teaching provides for one's own learning have long been recognized, tutoring associated improvements in measures like achievement for the tutor role than the lecturer role. However, scientists had confirmed that human teaching robots has been concerned with learning for the benefit of the robot rather than that of the human.

Intelligent autonomous robots are making their way into an increasing range of application areas: Manufacturing, transportation, surveillance and monitoring, rehabilitation, agriculture, service. Hence, if teaching robots can be confirmed to assist teachers to teach students in any schools. It seems that if can be applied to teach manufacturing workers how to produce any products in manufacturing process; it can be applied to teach drivers to drive cars, or teach pilots to control plane engineering machines to fly to sky. So, it can be applied to teach any occupation learners to learn how to operate any engineering machines to replace any human machine training teachers in possible. Hence, it is possible, future one day, robots can be any mahine operating occupation job trainees' teachers (tutors) or trainers.

In education industry aspect, for computer subject example, robotic systems have potential to increase studetn involvement and motivation and to improve the effectiveness of learning. Possible ways in which they could do so range from small personal robots used as teaching tools for areas like programming and computational thinking to move science-fictional future scenanios wirh humanoid robots supplement or even replacing teacher role in schools.

Recently " teachable agents" have been used as a tool to help students learn, e.g. s student is tasked with training a simulated computer agent on course material, which can lead to a deeper and more committed understanding on the student's own part. For example, a robotic presence in classrooms may one day help to solve problems of teacher shortage, as learning-by-teaching systems have historically been used to do. And the data about human teaching styles that will be generated by widespread use of teachable robots in classroom settings will be invaluable in developing approaches to help robots learn more effectively from humans, where it is rapid and natural instruction of a robot in a new task.

Consequently, whether future robot role is teacher role better or teacher's assistant role or tutor role better in education industry. It depends on the general student's preferable teaching choice in the school. Such as if the school's students prefer to be applied to be taught by human teachers. Then, robots can be choosed to be the school teachers' teaching assistant role to assist the school's teachers' teaching jobs only. Otherwise, if the school students prefer to be accepted to be taught by robots. Then, robots can be choosed to be the school teachers' role to replace human teachers to teach students in the school. Thus, future robots can be either teacher role or teacher's assistant or tutor role for child age, young age or adult age student learning consumers in any primary or secondary or university in future robot computer educational machine development.

Reference

Chou, C. Chan T., & Lin, C. 2003 Redefining the learning companion. The past, present and
future of educational gents, computers & education, v.40 n.3, pp. 255-269. April 2003.

Freeman, R. 2000, What is an intelligent tutoring system? Published in Intelligence 11(3): pp.15-16.

IFR, Statistical Department, World Robotics Survey, 2008.

Knox, W.B. Breazeal , C.Stone,.O: Learning from feedback on actions part and intended : In proceedings of 7th ACM/ZEEE International conference on human-robot interaction, late- breaking reports session (HRI 2012).

M. Mataric, " Reinforcement learning in the multi- robot domain," Autonomous robots, vol.4, no 1, pp.73-83. 1997.

Wenger, E. 1987. Artificial intelligence and tutoring systems. Los altos, CA: Motgan Kaufmann.

Artificial intelligence and the future of defense

Nowadays, artificial intelligence (AI) is widely knowledge to be one kind of the dramatic technology. However, it is expected to continue, to have a disruptive impact on human's private and public life, so defense and security will be no exception. But how exactly will these be affected ? How will (AI) defense and security is incremental in nature?

To research why artificial intelligence (AI) has possible to be used to cause autonomous weapons by human. We need to understand these three aspects of relationship. They include cybersecurity and artificial intelligence and machine learning and autonomous weapon systems relationship between of them.

Firstly, we need to know what is the mean of artificial intelligence and cyber defense/offense? It means defense of critical networks: real time, pattern finding, anomaly seeking, it must utilize machine (AI) learning algorithms to efficiently, and instantaneously respond to potential network threats as well as it means human on or out of the loop. On the loop : it means anomaly detection: human notified, IT analysis, response. Out of the loop: it means anomaly detection: (AI) decides best method of response: quarantine, honey pot monitoring, hack-back. Thus, it is possible that (AI) can be used , such as autonomous cyber weapon.

What is artificial intelligence and autonomous weapons? Autonomous weapons mean one kind of weapon that can be selected and engaged a target, without intervention by a human operator. Are these machines artificially intelligent? I believe the answer is not, because present weapons systems are not capable of human level reasoning. But, (AI) algorithms are presently employed to process sensor data, monitor system health, take and respond to vocal commands manage data, navigate. This, future autonomous weapons systems will require stronger (AI) to be secure and operationally and cost effective. Moreover, self-aware autonomous cyber systems are crucial.

What is cybersecurity mean? It means the ability to control access to networked systems and the information they contain. It is acted to prevent , detect, recover, react. It is application objects concern people, process, technology and it's application goals are confidentiality, integrity and popular availability. Thus, what is cyber weapon mean? Walware means viruses, Trojans, zero-days, worms ransomware, spyware etc. Does it require a particular objective? E.g. military paramilitary or intelligence. Does it require physical harm? E.g. functional harm or interruption? Mental harm? Is (AI) a technological weapon that it is an object or tool? What about when it is an weapon agent?

In simplicity, (AI) can be one of scientific weapons platform. When one day, it is invented to be applied to control war planes to fly to any countries to attack enemies or it is invented to be seemed to human to replace soldiers to bring guns or any weapons go to other countries to attack. So, it is possible that future any war defense planes, (AI) technological automatic control weapon can be replaced of human soldiers or war plane pilots to control any war defense planes to go to different enemy countries to attack them easily. It is very horror matter to threaten global human's ourselves life in the future , if (AI) automatic control war defense planes or (AI) automatic control machine soldiers were invented successfully.

Hence , when (AI) can be applied to weapons platforms, it structures that launch weapons, i.e. jets, ships, vehicles. (AI) platform and weapon and software architecture components are be done one (AI) technological weapons systems. Thus, human will encounter any (AI) benefits or risks (threats) causes in the same time as soon as possible. If we can predict when (AI) weapon system will be manufactured or invented successfully. Then, we can reduce (AI) weapon systems risks , if we can threaten any (AI) scientists continue to invent any undiscovered (AI) weapons in any time to avoid the future first time (AI) weapon war occurrence in possible.

The (AI) weapon system risk means autonomy: the ability to problem solve technological war , when (AI) weapon system is manufactured successfully, the power to act, how to damage the (AI) weapon system. The power to chance to stop (AI) weapon system manufacturing processes, ability to create a new goals, how to change the (AI) weapon system inventors' or scientists' minds to avoid to apply (AI) tools to achieve attack goals to change to another positive goal. Due to human can't know a prior what an autonomous (AI) weapon system will do.

Although, human is known what (AI) is , but human is also known when (AI) scientists whose emergent behaviors will do to change to do any negative behaviors from positive behaviors. Whatever (AI) weapon system design we use, there will be cybersecurity, problems arising from computation design/ complexity. Due to any one (AI) scientist can manipulate the system to act against itself, or who can utilize traditional " cyber weapons" against the (AI) weapon system, or who can manipulate the system to lie to humans, but also due to complexity, there is no way to know if it is lying or not or bounded rationality : satisficing.

Finally, the most serious (AI) technological invention risks are human is unknown these aspects of (AI) absolutely: They are not simple automatic systems, learning reasoning, communication of " self-aware" systems. Thus, human will face (AI) technological invention risks or threats. We need to find any methods to avoid (AI) weapon system is manufactured successfully to avoid (AI) technological war can occur in future anyone day.

(AI) system immoral intention

Why (AI) system can be invented to damage our society ? IS it possible to achieve this (AI) damage system successfully? ON (AI) attribution hand, it can be applied to cars, aircraft, which are subject to regulation designed to protect the public from harm and ensure fairness in economic competition. Thus, (AI) safety issue is important to scientists to consider.

IN general, the approach to regulation of (AI)-enabled products protect public safety issue should be informed by assessment of the aspects of risk that the addition of (AI) way reduce any respects of risk that it may increase. Also, where regulatory responses to the addition of (AI) threaten to increase the cost of compliance, or slow the development or adoption of beneficial innovations, policymakers should consider how those responses could be adjusted to lower costs and barriers to innovation without adversely impacting safety or market fairness.

For example, regulatory challenges that (AI) enabled present are found in the cases of automated vehicles. (AI)s, such as self-driving cars and (AI)-equipped unmanned aircraft systems. IN the long run, self-driving cars will likely save many lives by reducing driver error and increasing personal mobility, it will offer many economic benefits. Thus, public safety must be protected as these technologies are tested and begin to mature. Creating safe spaces and test beds for experimentation , and working with industry and civil society to evolve performance based regulations that will enable more uses as evidence of safe operation accumulates. Thus, it implies that any scientists can also invent (AI) system to control weapon defense planes or (AI) automatic machine human to do any soldier's behaviors to attack to any countries easily, instead of none driver automatic control vehicle invention. Thus, (AI) system can be applied to harm to human or achieve to damage our society aim by ourselves in possible.

The rapid growth of (AI) has dramatically increased the need for people with relevant skills to support and advance the field. AN (AI) –enables would demand a data literate citizenry that is able to read, use, interpret and communicate about data and participate in policy debates about matters affected by (AI). Thus, if (AI) technology is applied to assist human's social development and raising life enjoyment or benefits. It will bring positive impact to influence human's future life. Otherwise, if (AI) technology is unsafe to be applied to threaten human's society. It will bring negative impact to influence human's future life. Thus, (AI) scientists need to consider how to apply (AI) technology.

As (AI) technologies move toward deployment, technical expects, policy analysts and ethicists have raised concerns about unintended, consequences of adoption. Use one (AI) to make consequential decisions about people, often replacing decisions made by human –driven bureaucratic processes, leads to concerns about how to ensure justice, fairness, and accountability, the same concerns of human's safety issue. Thus,)AI) expects have cautioned that there are challenges in trying to understand and predict the behaviors of advanced (AI) systems.

Use of (AI) to control physical-world equipment leads to concerns about safety, especially as systems are exposed to the full complexity of human environment. A major challenge in (AI) safety is building systems that can safety transition from the closed world of the laboratory into the outside open world, when unpredictable things can happen. Adapting to unforeseen situations are difficult necessary for safe operation. Experience in building other types of safety artificial systems and, such as aircraft, power plants, bridges and vehicles has much to teach (AI) practitioners about verification and validation, how to build a safety case for a technology, how to manage risks, and how to communicate with stakeholders about risk. The risk means the harm of human's safety of (AI) damage system control machine invention. Thus, any (AI) scientists need consider moral responsibility when who decide to invent what kind of (AI) system machine to aim to bring human's benefits or attribute to human's welfare intention.

Thus, (AI) products safe invention matter will need any scientists' considerations. Because , if (AI) any products are unsafe or harm human's invention in the manufacturing process, it will bring any human's life danger when the (AI) system damage tools are invented successfully and are provided weapons to humans to use to attack other countries easily. It will cause future global human (AI) technological war occurrence.

I shall recommend the solution is necessary of ethical training for (AI) practitioners and students. Ideally, every student learning (AI) , computer science, or data science would be exposed to curriculum and discussion on related ethics and security topics. However, ethics alone is not sufficient. Ethics can help practitioners understand their responsibilities to all stakeholders, but ethical training should be methods for deciding good intentions into practice by doing the technical work needed to prevent unacceptable or immoral (AI) invention outcomes.

Hence, global human needs to concern (AI) weapon system invention security issue. Nowadays, (AI) has important application is increasing role for both defensive and offensive cyber measures. Currently, designing and operating secure systems requires significant time and attention from experts.

Challenges issues are raised by the potential use of (AI) in weapon systems. The United States has incorporated autonomy in certain weapon systems for decades, allowing for greater precision in the use of weapons and safer, more humane military operations. Nonetheless, direct human control of weapon systems involves some risks and can raise legal and ethical questions concern (AI) manufacturing process intention.

The key to incorporating autonomous and semi-autonomous weapon system into American defense planning is to ensure that U.S. Government entities are always acting in accordance with international humanitarian law, taking appropriate steps to control , to develop standards related to the development and use of such weapon systems. The United States has activity participated in ongoing international discussion on Lethal autonomous weapon systems and anticipates continued robust international discussion of those potential weapons systems. Thus, (AI) scientists have responsibilities to manage the potential to be a major driver of economic growth and social progress only, their (AI) intentions are not the global dominance aims absolutely, if (AI) product industry , civil society, government and the public work together to support (AI) positive development of the technology with thoughtful attention to its potential and to managing its invention threat risks to avoid (AI) products to manufacture to be used weapon tools.

Finally, I recommend that as the technology of (AI) continues to develop, practitioners must ensure that (AI) enables systems are governable, that what their inventions need to be openness to let public to know clearly and understandable; that they can work effectively with people and that their operation will remain consistent with human values and aspirations. Researchers and practitioners have increased their attention to these challenges , and should continue to focus on their future any (AI) inventions.

Hence, (AI) safe system ought to be applied to solve the biggest challenges that society faces, such as mobility for the elderly and those with disabilities, smart buildings may save energy and reduce carbon emissions, precision medicine may extend life and increase quality of life, smarter government may solve citizens more quickly and precisely., better protect those at any immoral invention risk and save money.

Moreover, (AI) enhanced education may help teachers give every child on education that opens doors to a secure and fulfilling life. Thus, these are the future human's potential benefits if the (AI) technology is developed to its benefits and scientists ought avoid to manufacture (AI) tools to cause weapon risks and challenges.

Consequently, the main point is that how experts invent (AI) systems. (AI) systems ought not be advanced weapon systems, it doesn't seem to be thought similar human soldiers mind and behaviors. (AI) system ought be systems that think like humans. (e.g. cognitive architectures and neural networks), systems that act like humans (e.g. pass the test via natural language process, knowledge representation, automated reasoning, and learning), systems that think rationally , e.g. logic solvers, inference and optimization and systems that act rationally e.g. intelligence software agents and embodies robots that achieve goals via perception, planning reasoning, learning , communicating, decision-making and acting function.

In conclusion, it is horror (AI) scientists will invent (AI) systems to be owned human's (soldier's) mind and attack strategic behavior to attack other countries easily, who must need to consider (AI) system ought be invented to own scientists' creating mind and non manual assistance functions for positive attribution to human's society. I expect that (AI) system can only be invented to create human's welfare in our future.

(AI) soldier weapon ethical, social and
economic negative impact

In the future, how human can avoid (AI) technological ethical, social and economic negative impact. Scientists need to concern these questions: how to develop of a good (AI) society, how the role and responsibility of the government, the private sector, and the reserch community(including education), in pursuing such a development, whether how the recommendation to support , such a (AI) system development may be in need of improvement.

However, none appers to deliver a comprehensive explicit vision of the role that (AI) system should play in mature information societies. Thus, (AI) 's potential contribution to social good shoud include an in-depth plan for linking in a comprehensive socio-political design questions of responsibility of the different stakeholders, of cooperation between them and of sharable values to understand of a good (AI) positive impact society, not a bad (AI) negative impact society.

Thus, the notion of mature information societies is introduced to stree the importance of addressing the current ethical challenges that (AI) poses in a comprehensive fashion.

It seems (AI) wil invention will be human's moral societal consideration issue. It concerns our (AI) scientists' moral issue, how who invent (AI) system to apply to which kind aspects. IF (AI) system was one direction on war weapon tools to similar to soldier's personal mind or attacking behavior. Then, it will bring poor social safety and poor economy growth our world, due to (AI) scientists' moral is low level.

Thus, the developed country US (AI) technological leader needs to focuse on the impacts of (AI)-driven customatin on the US job market and economy. It represents three specific policy responses to the perceived impact of (AI) on the US economy. They include these three aspects such as: How to invest in and develop (AI) for its many benefits, how to educate and train Americans for the jobs of the future and how to aid workers in the transition and empower workers to ensure broadly shared growth.

The future of (AI) influenced cyber conflicts need more than just the application of current and past solutions in order to ensure security and stability of societies, and avoid risks of escalation. To achieve this end, efforts to regulate cyber conflicts require an in-depth understanding of this new phenomenon, identify the changes brought about by cyber conflicts and the information revoluation, and defines a set of shared values that will guide the stakeholders operating to avoid the international (AI) war occurrence. This becomes clear when considering for example, cyber deterrence. Deploying conventional (cold war) strategies to deter (AI)-influenced cyber conflicts proves highly problematic and the urgent need to foster and coordinate new solutions able to account for the any kinds of conflicts of the cyber demain and of mature information societies to avoid (AI) technological war occurrence in the future.

We hope that in the on-going international conversations and reviews, the US government with further specify how " (AI) system invention law" fit into their vision of the future of society in this case the future of (AI) technological war and conflicts. Hence, (AI) scientists need to concern ethical issues related to (AI), like fairness, accountability and social justice can be addressed through increasing needs. Such as: how the creation of a new body focused on robotics and related (AI) system development to avoid to intent to apply weapon tools to provide advice on the policy, legl and consumer protection issues arising in these fields should be considered.

How to achieve ethical training of (AI) staff and ethical education of the public is certainly important responsibility for (AI) tools ethical behavior and design to the private sector and the citizens :

of unique challenges that (AI) brings to society in terms in fairness, social equity and accountability are addresses. Thus, the development of the (AI) technology and defining good (AI) remains problematic. In particular, the US government's innovation driven approach to defining the potential, positive impact of (AI) shows that more could be done to ensure that the opportunities and advantages brought about by (AI) are shared by all society.

An initial on Robotics, based upon the ethical framework and guiding principles is proposed. It should be complementary to legislaton and comprise ethical codes of conduct for Robotics researchers and designers, codes for research ethics committees as well as licenses (rights and duties) for designers and users. Thus, (AI) robotics invention of safety issues is very important considertion to any (AI) inventions or researchers. Every country's government ought have legal guiding to control their robotics' manufacturing intention. If their robotics (AI) is applied to seem to be soldiers to attack other countries to threaten their people's safety. Then, those (AI) inventors or researchers need to be punished by law.

In conclusion, I believe (AI) technology will be applied to weapon, when it's technological development is nearly mature to able to learn human's mind to do any behavior. During (AI) technology reachs thie mature stage, I predict the (AI) weapon tool , e.g. (AI) soldiers will have chance to be caused. This (AI) invention mature stage has these characteristics such as:

When (AI) invetion reachs this mature stage, computers and robots will develop conscious, intelligent, personified minds. Further, information technology devices and (AI) systems will be implanted into humans, enhancing, psychological and behavioral abilities and allowing for direct communication with artificial intelligent minds. There will be both artificial intelligence (AI) and intelligence amplification (AI) in the relatively near future stage.

During the (AI) invention reachs this mature stage, these will be an ongoing mulit-faceted integration of information technologies and human life. Humans and information technology will cooperate. Humans will increasingly immerse their lives and minds in (AI) systems of technological intelligence and virtual reality. The distinction between humanity and technology will increasingly close dependence.

During the (AI) invention mature stage reachs that the environment will be infused with information technology, becoming animated, communicative and more intelligent. The destinction between the artificial and the natural will increasing close dependence.

During the (AI) invention mature stage will expand through virtual reality, simulated and virtual reality will increasingly into normal reality, e.g. the (AI) weapons is virtual reality to seem to be soldier weapon.

Finally, during the (AI) invention mature stage is as the global expression of the evolving human-technology integration a " world brain" and " world mind" will emerge on the earth. This psychophysical (AI) weapon system will enhance and enrich the capacities of both individual and collective cogniton. This (AI) weapon system is a potential starting point toward the evolution of a cosmic brain and cosmic mind.

Thus, it is possible that the workship raw data was a unique way in which (AI) could be weaponized to cause war, during the (AI) invention stage reachs the invention mature stage. However, (AI) weapon manufacturing factory will be built possibly. In the future, how will we defins and locate (AI) weapon factories. Especially, as these factories are no longer solely buildings , but a mil of virtual and substantially different facilities, particularly as it shifts from a physical assemly and development model to a distributed and flexible network. Needing minimal raw materials to develop (AI) weapons, the phsysical location of their (AI) factories could be anywhere and their identification from the outside, nearly impossible. Given the expanding uses for intelligent and super-intelligent (AI). How will we tell the different form a location that is manufacturing (AI) for the creation of weapons versus creating (AI) for an innovative new gaming platform?

In conclusion, human needs to consider every (AI) scientist's personal ethical or moral mind and research intention and (AI) system invention of (AI) weapon factories cause. During (AI) invention reachs the mature stage if human expects to avoid (AI) technological war occurrence in future one day. The technological development on autonomous military robots, ideally among relevant social groups and actors including human-rights, activists, researchers developers, engineers, philosophers, policy-makers, military authorities, lawyers, journalists and the publis need to consider when human has effort to invent autonomous military robots successfully in the future one day. Finally, some ambitious countries or dominant global countries must like to apply (AI) autonomous military robots to be machine soldiers more than human soldiers if (AI) technology had reached the mature stage. So, future (AI) autonomous military robots will be the next choice of weapon to follow nuclear weapon. If civilians were used as a human (AI) soldiers, the weapon simply ignored them and targeted anyway. This scenario highlighted the dangers of proliferation and quick replication of autonomous weapons. Unlike nuclear weapon, a piece of code for (AI) artificial intelligent soldier could be obtained on the black market and replicated at little cost and the hardware for this type of weapon doesn't require costly or hard to obtain components and materials. Thus, (AI) artificial intelligent soldiers can be manufactured many at cheaper cost. Otherwise, manufacturing one nuclear bomb weapon will spend too much cost. Hence , it is possible that (AI) artificial intelligent soldier will be future new technological weapon to follow nuclear bomb weapon. Hence, any country government needs to legislate to control any (AI) scientists' inventions whether they are attributed benefits or welfares to human or damage human's safety.

How does (AI) robots' brain invention influence our lives?

(AI) research modeling the human brain has developed important technologies, and has overcome significant barriers. How will (AI) affect humanity in the near future? How will (AI) change our lives and our societies? Is the evolution of (AI) to humanity, or it represent a threat?

On white collar workers (AI) job replacement aspect, University of Tokyo, Institute of informatics, lecturers who had attempted to do experiments to take (AI) exams over a two year period. The (AI) achieved standard scores of around 50 in each subject, exceeding the norms for humans attempting the tests. The (AI)'s results in subjects emphasizing memorization, such as world history and Japanese history subjects were comparatively high, and the results of the study suggested that an appropriate selection of subjects would give at an 80% chance of passing the entrance exams of 80% of Japan's private universities.

So, if (AI) is applies to human white collar workers' job duties aspect, at this level, if white collar workers were replaced by (AI) in the future, around 30% of current staff would be replaced. Whatever, the outcome, large companies will be represented with two choices. One choice will be to protect their employees, but as a result lose their international competitiveness. The latter choice will enable them to reduce the cost of general duties, financial management procedures, accounting etc. general administrative job duties of cost in offices. Hence, it seems that (AI) will be possible to be invented to own human's brain ability to do some mind jobs in future on day.

However, the method called " deep learning" must be developed to cause (AI) to match human's brain ability as well as these were dramatic advances in technologies, such as image recognition and voice recognition, which form the foundation for (AI). Nowadays, this new method called" deep learning" does not reach the matured and stagnated stage. It needs to wait human to continue to invent to let (AI) to match human brain to achieve 100% owning human's mind ability. Nowadays, (AI) industry product include cleaning robots, smart TVs and future (AI) product development market. It will include self-driving vehicles, drones, and nursing robots.

On (AI) weapons applied aspect, if (AI) can be invented to own human's brain judgement and analytical abilities. Then, it is possible that it can be applied to attack enemy to cause war effect. For example, if weapons such as missiles were equipped with (AI) in the future, they would become able to decide on their own targets. Hence, human needs to apply restrictions when necessary.

On (AI) applied to analyzing information collected technological aspect, nowadays, every one will use wearable terminals to connect to the internet to obtain various types of information as well as computers will collect and analyze information on people. Our lives will probably be more reliant on these internet technologies than they are on smartphones today. When, (AI) can match human brain to own mind ability.

Then, (AI) can be applied to do any analyzing information and collection job duties aspect to raise large information restoring and remembering efficiency. For example, (AI) will be generally used and will be extremely useful in analyzing the information collected from wearable devices and stored in the cloud. (AI) will enable wearable devices to be of real assistance in our lives offering their users more intelligent support.

Rather than allowing (AI) to develop on serves, as something separate from humanity. It will be more meaningful to encourage its development via wearable devices, situating it under the control of human intelligence. The intelligence of (AI) will increase rapidly in the future. If this increase in (AI) occurs under human control, enabling humans to increase their own abilities, then surely it will be possible for us to put up a degree of resistance to the opposite scenario, the domination of (AI) over humanity. Hence, if (AI) can be invented to remember and store and make analytical judgement to collect any information from internet. Then, it will bring the effect, such as large international organizations' (AI) internet storage robots can bear in mind factors, such as competitors' privacy or business secret information, such as the loss equality between people and threats to privacy that will be stolen form the owning (AI) storing internet information remembering robots.

Consequently, what is the effect of successful invention of (AI) matching human brain's mind ability? (AI) present computers are adequately able to reproduce the emotional, conceptual and intuitive abilities of humans. Because of this, it is important that we should envision potential future problems that may manifest when we consider how to employ wearable devices. It will be essential to enhance our technologies in order to ensure that we can use (AI) under human control.

However, when a goal has been set. (AI) will implement an appropriate means for its realization. (AI) will be need as a tool by human society. If the capacities of analytical and judgement mind abilities of (AI) brain exceed those of human brain, it is difficult to imagine the type of technological, then singularity is represented by the creation of an (AI) by another (AI). It is important that we rapidly and accurately predict these developments, when image recognition and other individual technologies are functioning at a high level. There will be a considerable matter in different sectors of (AI) industry development.

Today, however, machines have become able to decide for themselves what they will learn, making it difficult to copy human's mind ability. What we must consider when machines exceed humans and (AI) surpasses human capabilities. May technologies exceed human capabilities, cars are faster than humans, planes are able to fly. Consequently, it brings a question that human needs to consider: When does (AI) brain technology be invented to reach the most reasonable stage to be accepted or stopped by humanity?

What is artificial intelligence human brain invention?

A machine is likely to achieve the ability of a human brain. Does it a scientific story? Some scientists has predicted that a US$1,000 personal computer will match the computing speed and capacity of the human brain by around the year 2020 year. With human reverse engineering, human should have the software insights before 2030 year. it is possible that of machine intelligence and exotic new technology for faster and more powerful computational machines from cellular automata and DNA playing cheese game competition case example, it proves that (AI) had been invented to own human's analytical and judgement ability to exceed the best cheese game human player's brain analytical and judgement ability. Then, it seems that (AI) will have possible to be built machine brains to achieve the exceed level of human brain's analytical and judgement ability in the future one day.

Supposing we scan someone's brain and restate the resulting " mind file" into suitable computing medium. Will the entity that emerges from such an operation be conscious? How have advances in electronic communications changes power relationship? For electronic book publishing case example, a book that looks at the principles companies must adopt to meet the needs and desires of this new kind of client. So, such as paper book can be changed to electronic book for human to read. Why can't human brain be changed to (AI) machine brain to do human's analytical mind and behavioral mind of activities to replace to do any human's daily analytical and behavioral mind activities?

Over the next few decades, machine achieve super intelligence, human will encounter a dramatic phase. Will it be a "WALL" a barrier as conceptually the event of a black hole in space. Such as (AI) brain invention case, an " AI singularity" ruled super-intelligence AIs, or a gentler " surge" into a post human era of agelessness and super-intelligence brain. Will future technology, such as bio-engineered pathogens, self replicating nan robots, and super smart robots run and accelerate out of control, perhaps threatening the human race?

If one day, (AI) brain is invented to achieve agelessness possibility. It means human's brain will be old to lose mind and analytical ability when human's age is increasing. Otherwise, (AI) machine brain age won't lose mind and analytical ability, due to (AI) machine is no age increasing possibility. It is a machine brain. If (AI) machine brain can be built successfully. Scientists need to consider technological ethic matter, such as the challenge of guiding nanotechnology in a constructive direction, advances in nanotechnology and related advanced technologies can not be inevitable, any broad attempt to relinquish nanotechnology would interfere with the benefits. When actually making the dangers worse.

Keeping in mind that intelligence machines are already making their way into our blood stream. There are dozens of projects underway to create blood-stream based " biological micro electronic- system" (bio MES) with a wide range of diagnostic and therapeutic applications BioMEMS devices are being designed to intelligently pathogens and deliver medications in very precise ways. For example, a researcher at the University of Illinois at Chicago has created a ting capsule with pores measuring only seven nanometers. The pores let insulin out in a controlled manner, but prevent antibodies from invading the pancreatic Islet cells inside the capsule. These nano- engineered devices have cured rated with type I diabetes, and there is no reason that the same methodology would fail to work in humans. Similar systems could precisely deliver dopamine to the brain patients, provide blood-clotting factors for patients with hemophilia and deliver cancer drugs directly to tumor sites. A new design provides up to 20 substance-containing reservoirs that can release their cargo at programmed times and locations in the body.

Another brain health technological related invention case, such as Kensall Wise, a professor of electrical engineering at the University of Michigan, who has developed a tiny neural probe that can provide precise monitoring of the electrical activity of patients with neural disease. Future designs are expected to also deliver drugs to precise locations in the brain. Also, kazushi Ishiyama at Tohoku University in Japan has developed micro machines that use microscopic-cancer tumors.

A particularly innovative micro machine developed by Sandia National labs has actual micro teach with a jaw that opens and closes to trap individual cells and then implant them with substances, such as DNA, proteins or drugs. There are already at least four major scientific conferences on bio MES and other approaches to developing micro-and nano-scale machines to go into the body and bloodstream. All these inventions are related to how to apply machines to copy human's brain knowledge in order to achieve to do any human's brain functions.

Finally, for Freitas envisions micron-sized artificial platelets invention case example, who could achieve hemostasis (bleeding control) up to 1,000 times faster than biological platelets. Freitas describes nano-robotic microbivores (white blood cell replacement) that will download software to destroy specific infections hundreds of time faster than antibiotics, and that will be effective against all bacterial, and fungal infections with no limitations of drug resistance.

Consequently, such as above machine health scientific invention cases, there were many scientists had invented any health machines to apply drugs to transfer to human's brain to attempt to reduce human's disease causing risks, such as reducing cancer cell increasing number. Why it is no possible that scientists can attempt to invent (AI) brain which can own human's mind ability to judge or analyze any matters to give opinions in order to exceed human's judgement and analytical ability.

How can artificial intelligent brain satisfy to human beneficial and natural needs?

Nowadays, new scientists' most familiar form of this vision in our times is genetic engineering. Specifically, the prospect of designing better human beings by improving their biological systems of a small, serious and accomplished group of tailors in the field of artificial intelligence and robotics. Their goal is a simply new age of post-biological life, a world of intelligence without bodies, immortal identity without the limitations of disease, death and unfulfilled desire. If human can understand why this fate is presented as both necessary and desirable, human might understand modern science can help us to enter the good life and good society stage when (AI) brain is invented by scientists successfully in our future life.

How can (AI) beneficial brain satisfy to human natural need? For relatively recent example, similarly as a long term trend beginning with the first mechanical calculators, the evaluation of computing capacity increases in speed over time and decrease in cost. From biological evolution has been invented to influence human brain, an electronic chemical machine with a great, but finite number of computer neuron connections, the product of which we call mind or consciousness. As an electro-chemical machine, the brain obeys the laws of physics, all of its functions can be understood and duplicated. And since computers already operate at far faster speeds then the brain, they soon will rival or surpass the brain in their capacity to store and process information. When happens, the computer will at the vary least, be capable of responding to stimuli in ways that are indistinguishable for human responses. At that point, we would be justified in calling the machine intelligent, we would have the same evidence to call it conscious that human now have when giving such a label to any consciousness other than our own.

At the same time, the study of human brain will allow us to duplicate its functions in machine circuitry. Advances in brain imaging will allow us to " map out" brain functions, allowing individual minds to be duplicated in some combination of hardware and software. The result, will be a world that is remade and reconstructed at the atomic level through nanotechnology, a world whose organization will be shaped by an intelligence that surpasses all human comprehension.

Whether or not today's humans are willing or able to " download" their brains into machines, there will come a time when all human beings will be intelligent machines in the future. Computer hardware will continue to get faster, cheaper and more powerful computer software will increase in sophistication. Brain research will continue to explore the " mechanics" of consciousness. Nanotechnology will continue to develop.

There are powerful incentives, commercial, military, medical and intellectual that will drive many of the advances that the extinctions desire if for very different reasons. Much of the work in artificial intelligence and robotics is open to the same defense that is made on behalf of biotechnology: If we don't do it, they will and why suffer or be unhappy when some new agent or invention is available that will solve the problem.

Finally, we already accept significant artificial argumentation and replacement of natural body parts when those parts are missing or defective. Over time indistinguishable from or " superior" to their biological counterparts as they employ increasing computer processing power. There are powerful incentives, commercial, military, medical and intellectual that will drive many of the advances that the extinctions desire, if for very different reasons. Much of the work in artificial intelligence and robotics is open to the same defense that is made on behalf of biotechnological if we don't do it, they will and why suffer or be unhappy, when some new agent or invention is available. That will cure the problem.

Finally, we already accept significant artificial augmentation and replacement of natural body parts when those parts are missing or deductive. Over time, such replacements are only likely to get more useful and perhaps eventually indistinguishable from or superior to their biological counterparts, as they employ increasing computer processing power. Nor is there an obvious distinction between using manufactured chemicals to fight disease and using " smart" nanotechnology. The extinction project is begun by offering new routes to fulfilling old promises about doing good for human beings. But, it doesn't necessary end.

In connection with machine intelligence, it does not seem very promising to try to limit the power or ability of computers. The danger (or promise) that computers might develop characteristics that lead some people to call them conscious and that this age of intelligent machines would mean our extinction seems remote when compared with their practical benefits. We already rely so heavily on computers that the incentives to make them easier to use and more powerful are very great. Computers already do a great many things better than we can, and there seems to be no natural place to enforce a stopping point to further abilities. Certainly mechanistic and reductionist assumptions about society, ethics and psychology the notion that we are atoms or animals, driven by chance or instinct, run deep in the present world.

Artificial intelligent brain future
innovation and attribution

16.1 (AI) brain invention successful factors

(AI) brain will bring what attribution to influence human's positive impact. How (AI) brain will be invented to apply to any businesses' needs. By how much the (AI) brains might exceed us remain unknown, but it could potentially be by a very significant degree. Future (AI) brain invention will have noted similar growth in everything from hard-drive storage density to the price and speed of DNA sequencing. On key feature of this technological growth that has not been adequately measured in the degree to which technology is becoming more intelligent.

16.1.1 (AI) brain test experiment

When there is an intuitive sense that (AI) programs/brains today are more capable than those of age, and that those were considerably " smarter" than the serial instructions that passed through the first supercomputers. Scientists will carry on testing (AI) to do any experiments to improve (AI) brain development. They will assess the progress of artificial (general) intelligence, but the need for intelligence tests and tests for other cognitive abilities will be tested in the forthcoming decades for bots, robots, avators, " animats" etc. and any collective system of these and biological systems (humans and non-human animals).

When (AI) brain technology is invented successfully, the idea of a super intelligent computer means invention successfully also. Whether super intelligent computer or (AI) brain can be invented successfully. Similarly, many think that the future beyond the technological singularity is unknowable or even unimaginable.

Since its conception, the idea has been examined and explored by technologists. (AI) brain invention will be mean human-equivalent (AI) vs human-level(AI). Many machine intelligence tests is the exclusive focus on identifying systems that achieve human equivalence. Considering our experience with studying non-human animal intelligence.

The need to distinguish human equivalent (AI) from human level (AI) seems critical. Perfect human intelligence , such as (AI) brain invention is likely to be extremely difficult to achieve. Potentially every aspect of the biological processes involved would need to be translated with very high fidelity.

Human level (AI), such as (AI) brain invention is another matter. Achieving capabilities that are equivalent to those of the human mind could be very feasible if it is not limited to perfectly processes involves. For instance, some pattern recognition algorithms are already superior to human abilities. This machine ability is not achieved by duplicating the processes our brains use, through some of the methods have been inspired by them. If researchers had been limited to replicating the brain's mind processes, We would still be waiting for the development of a (AI) brain machine equivalent.

Tests that would seen to have a reasonable chance of successfully testing non-human intelligence are those that use mathematics to define the value of a given challenges for (AI) brain invention. Such as Capability-test has some potential to generate meaningful data about (AI) machine intelligence, e.g. others in complexity theory test, it presents a series of abduction and prediction problems, similar to those in standard IQ tests to (AI) brain. Hence, IQ test and C-test will be potential tests to test (AI) brain ability.

(AI) brain test goals include that to identify a range of possible types of mind such as: super-fast human mind, mind with operational access to its source code, any mind capable of general intelligence and self awareness, general intelligence without self-awareness, self-awareness without general intelligence, super-logic , machine without emotion, mind capable of imaging greater mind or creating greater mind to compare human's brain abilities. Thus, any IQ or

Capability test experiments aim to evaluate whether (AI) brain mind ability which can exceed to human brain mind ability. Thus, if future one day, scientists could prove (AI) brain mind ability can exceed to human brain mind ability. Then, they believe (AI) brain invention has ensured to achieve success.

16.1.2 (AI) brain invention successful factors

Hence, scientists want to invent (AI) brain successfully. They need to solve these challenges. Such as How robots and computers have progressively supplemented humans, initially only in relatively simple computational and manipulation tasks, but more recently in higher cognitive tasks that used to be the pre negative of the human brain, including language, mathematics, probabilistic reasoning and decision making.

An important challenge is how to enhance the productive interactions between humans and artificial intelligence? The important challenge include these major successful factors to invent (AI) brain technology, such as:

What is the state of the art in (AI) software and machine learning?

Can all aspects of brain function be manufactured by artificial system?

What is the proper form of mathematics that may capture the operation of minds and brains?

How to make (AI) brain feels consciousness?

What would it be taken for a machine to pose a sense of art in (AI) brain software and (AI) brain machine learning by artificial system?

Will machines soon surpass us in all domains of human competence?

What is the proper form of mathematic that may capture the operation of minds and brains?

What is consciousness to (AI) brain invention?

Could a machine be endowed with an artificial consciousness?

What would it take for a machine to posses a sense of self?

Will intelligence machines soon pose a danger to humanity of (AI) brain is invented to reach the mature stage successfully?

Is it possible to design and construct an intelligent robot with an artificial brain sense of ethics?
How can we enhance the humanitarian uses of artificial intelligence brain and owning mind ability of robotics, in particular in the field of education, health and emergencies?
Consequently, above all these challenges, I recommend scientists need to solve to achieve (AI) brain invention in order to reach (AI) brain invention mature stage easily.

16.2 Artificial intelligence brain invention opportunities and challenges

If scientists focus wrong direction to invent (AI) brain, it will bring wrong marketing development to attribute any benefits to human. So, they need to reduce a mismatch of timescales between the pace of commercial innovation and (AI) brain invention process, reduce an underappreciation of the fundamental unpredictability of (AI) brain autonomous systems and reduce a lack of a university agreed upon conceptual framework for any (AI) brain invention and reduce a disconnect between the (AI) brain design of any kind of (AI) autonomous robots. Thus, scientists need examine these gapes, provide a roadmap of opportunities and challenges and identify areas of any (AI) beneficial functions to be attributed to human to use for our daily life needs.
Future (AI) brain market opportunities, it is rapidly growing innovations in digital -electronic and information technology had development of new intelligence, surveillance, and reconnaissance platform and battle management capabilities, precision-strike weapons, stealth aircraft, smart weapons and sensors and tactical exploitation of space (e.g. GPS).
Any one of these aspects will be (AI) brain future marketing development opportunities. Future (AI) brain invention of so called " narrow AI", (i.e. non-sentient artificial intelligence, whose problem-solving capability is confined to one narrow task. For example, (AI) brain needs to find the best method to win any one of chess player

in any both human and (AI) robot chess playing game.

In smart weapon strategy industry, (AI) brain is needed to design how to analyze or mind to protect whose country to avoid enemy attack in any war by the best weapon protection strategy. In education industry, (AI) brain is needed to design how to analyze or mind how to assist teachers to educate whose students by the best education method. In aircraft industry, how to design (AI) brain to analyze or mind to assist pilot to make the most correct flying direction judgement to fly in the most safe way. In space exploitation industry, (AI) brain is needed to design how to find undiscovered natural resources to supply to human to use in anywhere space.

Thus, future (AI) brain invention needs have these features/ characteristics to be designed. They include: (AI) learned on its own, where to find the information it needs to accomplish a specific task, (AI) can predict the immediate future from studying any matter, (AI) automatically needs to be inferred the rules that govern the behavior of individual robots within a robotic swarm simply by watching, (AI) needs to be learned how to navigation the acquired memories and experiences, much like a human brain, (AI) speech recognition needs to be reach human parity in conversational speech, (AI) communication system needs to be invented its own encryption scheme, without being taught specific cryptographic algorithms (and without revealing to researchers how its method works, (AI) translation algorithm needs to be invented to remember fluent language to more effectively translate between any two languages (without being taught to do so by humans), (AI) brain system interacted with its environment (via virtual environment) to learn and solve problems in the same ways that a human child can do, (AI) based medical diagnosis system needs to be achieved 99% percent accuracy in any medical reviewing researches (at a rate minimum 30 times faster than humans), (AI) poker playing program brain development needs to be defeated some of the world's best human poker players during a minimum three-week-long tour, (AI) brain development needs to be effectively " read

minds" of human test subjects looking at pictures of faces, via functional magnetic reasonable images of brain activity.
Consequently, (AI) future brain development needs to follow above directions to be invented to attribute to human's satisfactory needs.

16.2.1 (AI) brain legal remembering attribution

Future, (AI) robots can assist lawyers to deal any legal cases more efficient. If human understands that smart (AI) technology is not to replace human lawyers, but to make a better lawyer that forces who to use, emotional intelligence, and capital on lawyers' mind, then human lawyer have made the first step in future, proofing whose legal service business from (AI) legal robots' assistance. (AI) promises to be a real advantage for today and tomorrow' lawyers having to deal with the rate of legislative evolution and technological change.

In future, for the better with regard to technology in any law firms. According to the ALM 205 law tech. survey 95% of firm leaders and technologist respondents agreed with recent decisions by management regarding the firm's technology in legal service profession.

How can (AI) robots be applied to legal service industry by legal service firms? The (AI) reality in the legal world, it includes in relation to the four key elements of legal service provision, such as commodity, research, reasoning and judgement, exist and can support or replace certain aspects of every lawyer individual jobs both fee earning processes and business processes can be supported and/or replaced by expert systems, cognitive computing, robotics automated systems, (AI) and the machine learning, clients demand and expect more speedy, accurate, expert, creative, intuitive and accessible legal advice, it can assist young lawyers to innovate / tech. focused firm, reducing pressure both from within law firms (or in house teams) and from clients to respond to the demand for client-designed service from (AI) robots assistance, technology related projects that are both user and client -centric need to be implemented successfully.

Due to the deployment of (AI) robots in the legal ecosystem where lawyers, firms, general counsel and clients are beginning to believe (AI) capability technologies, (AI) robots can assist lawyers to increasingly become more productive, efficient, accurate, better quality, less labor intensive and time intensive and the role of the lawyer is gradually changing. If we break down a lawyers' tasks in a legal project from beginning that can handle the majority of these four tasks far more quickly and accurately than human lawyer. (AI) robots can handle legal task in the four aspects as below:

First in legal aspect, it can be used for deep research and processing, such as extracting specific pieces of information from land registry documents, (AI) technology is placed top of a document set including client guidelines and similar forms. It searches through documents and extracts key data points to provide a report of data for improved coordination with clients.

Second on managed services technology aspect, (AI) platform which could have a huge advantage for general counsel and law departments in corporations and for clients of all company sizes.

Third on reading aspect, (AI) brain program that reads and analyzes , e.g. clauses in loan agreements. Its program helps its lawyers through transactions and points. Then toward the correct precedents of each stage of a process.

Fourth, on academics aspect, (AI) robots can assess the merits of personal injury cases. It can automatically review high volumes of contract documents to identify provisions that could potentially be impacted by contract law regulation.

Thus, all of these systems can handle large quantities of structured and unstructured data, and assist with the process management, research, and reasoning elements related to legal issues. Thus, in future, (AI) brain development can be invented to apply to knowledge research job, such as legal industry.

Artificial legal intelligence presents a thought-provoking approach to both computational models of legal reasoning and the use of evolutionary thinking about the law. The visions of computerized artificial legal intelligence , a vision of developments in both technology and legal history. A number of creative research projects have applied artificial intelligence techniques to the domain of legal reasoning.

Consequently, (AI) brain for legal industry marketing development ought concentrate on those three fields of artificial intelligence at most relevant to work in the legal areas in order to achieve the excellent attribution. Such as case-based reasoning, expert systems and neural networks. Artificial intelligence program, such as the legal reasoning programs. Thus, (AI) brain invention needs to own these human legal concept knowledge in order to achieve (AI) legal brain program development successfully.

16.2.2 Whether human mind can create to (AI) brain mind

How can (AI) scientists build a machine that think? In fact, there was general agreement that minds can be existence on non-biological substrates and that algorithms are of central importance to the existence of minds. However, there are much debate about the raw hardware power present in organic brains, such as (AI) brain.

I think (AI) scientists need to invent powerful hand ware, e.g. commercial digital signal processing might be, giving an appearance even to digital operations, but nothing would ever make up the intellectual runaway that is the essence of the singularity.

I also think raw hardware power is not be able to organize the parts to behave in a super-human way as well as I also think powerful software complexity is the main factor to solve (AI) brain mind invention challenge.

Hence, future super-human (AI) brain invention will ought consider how to invent superhuman software more than hardware, because software can store any memory, i.e. human mind. When, (AI) brain invention which can achieve to own human mind ability. Then, (AI) scientists need to consider ethic matter: Does the future of (AI) pose an existential threat to humanity? How do we present learning algorithms from morally objectionable biases? Should autonomous (AI) be used to kill in warfare? How should (AI) systems be in our social relations? Is it permissible to fall in love with an (AI) system? What sort of ethical rules should (AI) like a self-driving car use? Can (AI) systems suffer moral harms? All those ethic matters. I think

(AI) scientists need to consider after (AI) brain owns human's mind ability because it is possible that (AI) robots will harm human if they are educated to do any wrong or illegal or immoral mind ability. Thus, (AI) scientists need to consider (AI) robot's moral mind and judgement behavior.

16.4.3 Brain-inspired intelligent robotics

How to solve fundamental problems in the areas of brain sciences and brain-inspired intelligence technology? One way scientists seek to accomplished the mission is to develop brain-inspired hardware including intelligent devices, chips, robotic systems and brain inspired computing systems. (AI) scientists ultimate goal, but since the robot's memory and learning after the human brain, there is still much to learn about neurobiology before that goal is attached.

In (AI) brain university research aspect, two schools of thought have emerged in robotics: bio-logically robots that include a body, sensor and actuators, and brain-inspired computing robot.

Robots have found increasing applications in industry, service and medicine, due in large part to advances achieved in robotics research over the past decades, such as the ability to accomplish complex manipulations that are essential for automated product assembly. However, robots still have these weaknesses which need to be solved if (AI) scientists expect (AI) brain invention can be success. Robots still lack truly flexible movement, have limited intellectual perception and control, and not yet able to carry out natural interactions with human. These deficits are especially critical in service robots. A critical concern of government, academia, and industry is how to advance research and development for the key technologies that can bring about the next generation of robots. Developing robots with more flexible manipulation, improved learning ability and increased intellectual perception will achieve the main goals for (AI) scientists' solutions.

Consequently future (AI) brain -inspired intelligent robotic invention needs have these competitive or attractive abilities or strengths to compare computer storage ability. Such as (AI) brain needs have perception to exceed computation ability, adaptation exceeds computing speed, flexibility exceeds, computer memory access speed, cognition exceeds computer memory lifetime, learning exceeds computer memory capacity and innovation exceeds computer memory storage abilities. Hence, (AI) brain innovation must need to exceed general computer storage, lifetime, capacity, learning abilities if (AI) scientists expect (AI) robots will be popular to be applied by any service, manufacturing, education industries.

16.3 How can web intelligence need (AI) brain informatics?

Brain informatics (BI) invention will be a new inter-disciplinary field that systematically studies the mechanisms of human information processing from both the macro and micro view points by combining experimental cognitive neuroscience with advanced information technology. (BI) studies human brain from the viewpoint of informatics (i.e. human brain is an information processing system) and uses informatics (i.e. WI centric information technology) to support brain science study. It seems that (AI) scientists need to further understand how human intelligence and brain sciences development through brain sciences fosters innovative web intelligence research and development because innovative web intelligence research will have ability to assist (AI) scientists to research how to invent (AI) brain robotic technology more easily. The synergy between (web informatics) (WI) and (brain informatics) (BI) advances our ways of analyzing and understanding of data, knowledge, intelligence, and wisdom, as well as their interrelationship, organizations and creation processes. Web intelligence is becoming a central field that information technologies and artificial intelligence to achieve human level web intelligence.

(WI) may be viewed as applying results from existing disciplines , e.g. artificial intelligence (AI) and information technology (IT) to a totally new domain the world wide web. (WI) may be considered as an entrancement or an extension of (AI) and (IT), (WI) introduces

new problems and challenges to the established disciplines.
Thus, developing human-level web intelligence will be seem to develop brain level artificial intelligent technology. Because brain informatics (BI) is an interdisciplinary field to systematically investigate human information processing mechanisms from both macro and micro points of view by cooperatively using experimental, computational, cognitive neuroscience, and advanced (WI), centric information technology. It attempts to understand human intelligence in depth, towards a holistic view at a long term, global vision to understand the principles and mechanisms of human information processing system (HIPS).
So, I recommend (AI) scientists needs to research how to invent brain informatics technology and web informatics technology to achieve how to apply (AI) brain to analyze and judge or mind web informatics ability to compete internet (web site) communication tool product industry. Hence, if (AI) brain can be applied to web informatics industry will increase attraction to an internet user market in the future.

16.4 Why model of sustainable development environment industry be (AI) robot attractive market

In the world scientific literature various conceptions of sustainable development are found, however, their basis are formed by three dimensions: environmental, economic and social development. The biggest attention is concerned to the environmental dimension which forms the basis of existence of social environment and economy.
The reason is emphasized that each person must preserve and manage natural resources as the basic of economic and social development. However, in the sustainable development strategy, the sustainable development is understood as among environment protection, economic and social society that form the basis to achieve the universal welfare for present and future generations, it is difficult to let human to adapt to live in pollution environment. Thus, environment pollution will be human's concerning issue. It

implies how to protect environment pollution which will be human's future challenge. If (AI) brain invention can be applied to how to solve environment pollution challenge. It is very attractive attribution to human (AI) brain environment protection function can be invented to include such as : how to apply (AI) brain to do mind to give opinions or do any environment protection behaviors/ activities to assist human how to effective use of natural resources, how to effective use of universal economic society's welfare, do strong social guarantees during the period of strategy's implementation (until 2020 year to achieve global (AI) robots environment protection mission from (AI) robots' behavioral assistance. Because environment pollution will influence world's climate to be worse to cause our food or vegetable can not grow up easily. Then, human will face food / vegetable shortage challenge. If (AI) brain invention can be applied to solve environment pollution aspects, then human will avoid food / vegetable shortage crisis.

16.4.1 How to invent (AI) brain's environment protection ability?

All similar problems significantly promotes scientists to research for new technologies of data extraction progressive software of data extraction operating on the basis of artificial neural networks allow finding the relations among various types of data in the huge data. Due to the data extraction technologies, it is possible to prove empiric observations to group, to process and model big amounts of data, distinguishing unknown schemes in data and using them in future activities (Rotman M. J., 1995).

It seems scientists believe gather every country's climate environment data to predict environment climate change will have chance to avoid environment pollution challenge. In addition, the traditional statistic methods applied to process digital data of the indicators of sustainable development environmental dimension are not able to analyze the present situation of environmental dimension in the context of sustainable development to present possible reasons of change of environmental dimension problems to generate future forecasts characterized by a high level of accuracy.

Consequently, if (AI) scientists can apply (WI) web informatics technology and (BI) brain informatics technology both to apply to (AI) brain technology to gather global climate environment daily change data. Then, I believe (AI) brain invention can assist human to solve future environment pollution challenge Then, it is possible that (AI) brain can attribute to environment protection or how to give opinions or choose to do any environment pollution activities to avoid global warming crisis.

16.5 (AI) Asia market

When (AI) brain is successful invention, I think Asia will be a new market to need (AI) brain attribution for Asia consumer needs. I believe Asia people expect (AI) can attribute to let them to use. The (AI) ability includes the ability of machines and systems to acquire and apply knowledge, and to carry out intelligent behavior.

This includes a variety of cognitive tasks (e.g. sensing, processing, oral language, reasoning, learning, making decision) and demonstrating an ability to move and manipulate objects according). So, future Asia people (AI) consumers expect (AI) robots can attribute to whose society's needs, such as: how to apply intelligent systems to use a combination of big data analytics, cloud computing, machine-to-machine communication and the internet of things (IOT) to operate and learn. Asia people expect (AI) robots can give beneficial attribution to them, e.g. talking or playing a game for Asia young entertainment market; (AI) robots need to reflected by physical substance (such as any talking or playing game robot player).

In this sense, (AI) is like a human brain. For Asia (AI) robot service industry need, Asia (AI) robot clients expect to use soft robotics (robotic process automation) can be used to meet Asia (AI) robot service consumers' expectation of automation repetitive tasks and common processor needs, such as client servicing and sales without the need to transform existing IT system maps (e.g. (AI) salespeople robots, or (AI) service robotic).

In (AI) office task aspect, Asia office consumers expect algorithmic game theory and computational social choice of (AI) robots to be attributed to replace some office human workers' tasks. Such as (AI) systems tat address the economic and social computing dimensions of (AI), such as how systems can handle potentially incentives, including self-interested human participants or firms, and the automated (AI) -based agents representing them, e.g. complex or simple office administrative tasks, e.g. typing, accounting, filing, etc. general office tasks which need human office workers who use computers to work to be replaced by (AI) robots to do. So, Asia office (AI) robots users who expect any office human administration tasks can be replaced by (AI) robots to do.

In Asia computer vision market, Asia computer vision (image analytics), users expect (AI) robots can be replaced to human computer image workers to shorten time to work, or raise image quality to be more clear in the process of pulling relevant information from an image or sets of images to advanced classification and analysis. Such as hospital or clinic x-ray image vision (AI) robot invention, photo image (AI) robot invention etc. any Asia image industry (AI) robot market need.

In Asia collaborative systems work with human (AI) robot market, Asia autonomous systems robots users who expect (AI) robots can be applied its models and collaborative systems to help them to develop autonomous systems that can work collaboratively with other systems and with humans, e.g. car manufacturing, computer manufacturing or any machine manufacturing products. So, Asia machine related manufacturing product industry manufacturers who expect to apply (AI) robots who can assist factory human manufacturing workers to manufacture any products in the short time efficiently.

In Asia language teaching (AI) robot market, language educators expect (AI) robots own natural language processing ability, algorithms that process human language input and convert it into understanding representation, such as Asia translation education market (PWC).

Thus, Asia language teaching businessmen expect (AI) brain invention which needs to be designed to own these above abilities of (AI) robot's manufacturers expectation to provide to them to use from any (AI) robots import.

16.6 Is the brain a good model
for machine intelligence?

The actions of a human " computer" using paper and pencil to perform a calculation (as the world meant), into a formalized machine, manipulating symbols on an infinite paper tape. But I believe it still has challenges to influence human brain can be invented to machine intelligence successfully.
I think (AI) scientists need to solve these further challenges to influence (AI) brain invention success. The limitations include such as below:
Computation is based on functions of integers is limited. They must let computer data can change to words, then words can change images, then any images can change to storage to let (AI) brain to remember to do any analytical and judgement able tasks to make any actions in the short time finally. It is (AI) brain behavioral and analytical mind speed limitation.
Moreover, anther limitation concerns biological systems clearly difference, they must respond to varied stimuli over long period of time, those responses any changes alter their environment and subsequent stimuli. The individual behaviors of social insects, for example, are affected by the structure of the home, they build, and the change their behaviors. Nowadays, (AI) brain invention is called computational neuro science, which have assured that the brain is a computer, it means a machine that is algorithms and architectures. Second, neuro science findings may validate the energy ability of existing algorithms being integral parts of a general (AI) system. It means (AI) robots change adapting environment limitation. It means how (AI) brains adapt to choose to make any analytical mind as well as how to be influenced by external environment to do their behaviors or actions in the efficient way, e.g. how to manufacture many cars efficiently in one factory in the short time.
Consequently, to solve these limitations, we need to know how to

apply correct conceptual knowledge to let (AI) robots to learn, e.g. for example, if we know how conceptual knowledge was formed from perceptual inputs, it would crucially allow for the meaning of symbols in an artificial language system to be grounded in sensory " reality". When (AI) scientists can achieve how to solve all above limitation challenges, then (AI) brain invention will achieve more easily.

Reference

PWC, Sizing the prize, see: https:// www. pwc. com/gx/en/issues/ data-and- analytics/publications/artificial-intelligence- study.html.

Rotman M. J., Data mining-a practical approach to database marketing (1995). IBM.

Rethinking morality in management science

What is morality in management science? Why does management need to concern moral behavior? Because most management researchers feel whose behaviors are possible right, but in moral view point, whose behaviors are possible wrong. This is a subjective personal moral judgement to management researcher individual behavior in any organization. However, if the management researchers can know how to judge whose behaviors whether ar more moral reasons to be accepted in societies . The, who can judge how to choose to do more right moral behavior more correctly in any organizations.

Some management scientists should not assume that one value can be objectively better than another, or that any values are objectively right or objectively wrong. So it brings this question: Are values objective? Are there sometime objective by god reasons to moral actions? Perhaps it is time to rethink hesitation about accepting moral objectivity in management search. Usually management researchers choose to decide to do immoral management behaviors. It is possible that who is seemed to have effort to help whose organization to gain benefits and dominant economy. So, in economic view point, the management researchers are helping their organizations to gain, it is possible that the management researchers do not feel who are doing immoral behaviors to assist their organizational development. Also, even whose behaviors are legal, but it is not represent their behaviors must be morality in societies.

Hence, knowing how to judge whether it is moral behaviors, it is important to every management research in nowadays organizational environment. In any organizations, that agency theory fails to predict corporate performance, because of mechanisms of monitoring, independence and incentives (Dalton et al. 2003: Dalton et al. 1999) . Due to this reason of self-interested, so management level staffs (researchers) often do immoral behaviors to influence whose whole organizations to bring negative effect to publicity. Because management researchers will feel themselves self-interest is important than whole organizational interest. It is self interested immoral behaviors, although it is not represent themselves behaviors must be illegal in society.

However, good explanations of moral / immoral behvior and how management researchers are affected by moral principles are ones that sometimes make use of the objective structure of moral reasoning. This relates to equate what ought to be with what " is". But is is to say that because management researchers think in ways about ethics, those patterns may be revealed in their actual behavior. Take a simple example related the value of fairness, e.g. the management researcher feels unfair compensation or welfare treatment or salary is provided from whose employer. So, it influence who choose to do immoral activities in whose organizations. So, it seems unfairness is a factor to cause management researcher's immoral behavior or the another factor of perhaps management researchers' hesitation to refer to moral principles when explaining behavior is from the assumption that moral reasons or principles can never serve as causal explanations of behavior.

Anyway , in any organizations, permission and obligation in ethics necessity and possibility in logic principle is essential to be prohibited to every management researcher behavior. The reason is because just or unjust (fair or unfair) behaviors are caused by economical self-interest of human behavior (management researcher behavior). Also, the prior organizational fair researchers are mostly explored relationships between fair or unfair behaviors in the one hand, and self-seeking motives on the other hand.

Consequently, I conclude that general management researchers choose to decide to do any kind of immoral behaviours in organizations. The reasons are usually due to that who feel unfair welfare or compansation treatment or economic self-interest motive factor to influence tham to choose to do unethic or immoral behaviors in their organizations. Thus, employers need to consider how organizational policies are applied prohibit management researchers choose to do any unethic or immoral behaviors in any suitations in organizational environment easily.

3.1 How common morality relates to business and professions?

Artificial intelligent scientist moralty

Supposing you are one mathematician/computer scientist in first mass computer laboratory. You need to design the root language and operate systems for the different computer languages. The first mass produced commercially available computer in world . Besides writing programs, you was a true pioneer in designing and developing whole languages for computers including the most successful languages for computers general purpose. You need to concern these issues:

How is a computer different then a calculator?
What is a computer language?
How does a computer language work?
How does it let us talk to machine?
How can you create a computer language?

Jurisprudence and mathematics are often grouped with the sciences. Some of the greatest physicists have also been creative mathematicians and lawyers. There is a continuum from the most theoretical to the most empirical scientists with no distinct boundaries. In terms of personality, interests, training and professional activity, there is little difference between applied mathematicians and theoretical physicists.

Computer ethics is a part of practical philosophy which concerns with how computing professionals should make decisions regarding professional and social conduct.

Computer morality can include such as:

a. The individual's own personal code.

b. Any informal code of ethical conduct
that exists in the work place.

c. Exposure to formal codes of ethics.

To understand the foundation of computer ethics, it is important to look into the different schools of biology ethical theory. Each school of ethics influences a situation in a certain direction and pushes the final outcome of ethical theory Relativism is the belief that there are no universal moral norms of right and wrong. In the school of relativistic ethical belief, ethicists divide it into three connected but different structures, subject (Moral) and culture (Anthropological). Moral relativism is the idea that each person decides what is right and wrong for them. Anthropological relativism is the concept of right and wrong is decided by a society's actual moral belief structure. Deontology is the belief that people's actions are to be guided by moral laws, and that these moral laws are universal. Utilitarianism is the belief that if an action is good it benefits someone and an action is bad if it harms someone.

This ethical belief can be broken down into two different schools, Act Utilitarianism and Rule Utilitarianism. Act Utilitarianism is the belief that an action is good if its overall effect is to produce more happiness than unhappiness. Rule Utilitarianism is the belief that we should adopt a moral rule and if followed by everybody, would lead to a greater level of overall happiness. Social contract is the concept that for a society to arise and maintain order, a morality based set of rules must be agreed upon. Social contract theory has influenced modern government and is heavily involved with societal law. Hence, you need have computer scientific morality to decide whether who are/is your new computer language invention's customer(s) group(s) and how who will apply your new computer language to influence society to bring harms or benefits more in the future.

I think moral issues concern the environment, specially the treatment of animals, and such issues as abortion and euthanasia, because these scientists often need to make moral decisions and judgement to make any experiments. The uncontroversial nature of those matters is shown by every scientist's lack of hesitancy in making negative moral judgements about whose who harm others simply because they do not like them. It is shown by the same lack of hesitancy in making moral judgements, unfair deception, breaking of promises, cheating, disobeying the law and not doing one's duty. For example, doctors choose animals' bodies to carry on killing any cancer cells experiments. It is immoral behavior to any animals, which have not dead. Because they used to be attacked by any kind of cancer cells and than doctors will give any new medicines to treat their bodies to attempt to test whether the new medicines are effective to treat any cancer cells. Do you feel animals won't feel hurt or painful when which bodies are tested to attack cancer cells by any new medicine experiments? If it had no any new medicines to be confirmed to attack their cancer cells in these animals' bodies successfully. Then, these animals will fell painful to die, due to any cancer cells experiments. For another example, whether these was a moral difference between a rational refusal of food and fluids by a competent refusal of medical treatment. To this patient's refusal of medical treatment suitation, the doctors need to do moral judgement whether who ought give medical treatment to the competent terminally ill patient or ought not give medical treatment to the competent terminally ill patent. It is too difficult to make moral judgement because the competent terminally ill patient who is probable to feel painful to alive. So, who decides not accept any medical treatment to reduce painful alive. But is the doctor promoted to him/her does not attempt to give any medical treatment to the competent terminally ill patient. It seems that the doctor shortens the patient's life days. Hence, the doctor will feel difficult to make right decision in the medical treatment case, also it is not absolute right or worng answer to the doctor's behavior.

In conclusion, I think any organizations need have rules to control their employees' behaviors, including professionals or general employees. So, the common morality, it includes: rules prohibiting acting or attempting to act in ways that cause, or significantly increase the prohibility of causing, any harms that all rational persons want avoidance. Whereas it is possible to obey the moral rules all of the time impartically with regard to with regard to everyone in our societies.

What is professional ethic? Professional ethics is not distinct from common morality. Rather, a profession takes on certain duties that are not duties for those duties must be compatible with the framework provided by common morality. Nor can any business impose duties on its employees that would be considered morally acceptable by all informed, impartial, rational persons. Also, a job can't impose duties that are morally unacceptable. For example, a driver of a getaway car does not have a duty to help bank robbers escape after a bank robbery, even if who has been paid to do so. An employee of an advertising company does not have a duty to help compose an advertisement that will persuade young people into smoking ot taking any other additive drug. Scientists who are employed by a cigarette company do not have a duty to help make that producing more addictive.

Consequently, it implies professional ethic is more strict to compare to general moral behaviors to general employees behaviors in any organizations. Because professionals need have better ethic to conduct daily behaviors for whose jobs and duties to compare to general employees in organizations. Finally, it also explains that it must have the difference between general morality and professional ethic in our societies nowadays.

However, (AI) technological scientists and other kind of scientists who need have moral or ethic consideration because (AI) consumers and any science product consumers who must concern their products whether their products will bring either safe use or unsafe use from law regulation prohibition.

Reference

Beauchamp T.L. Chikdress & J.F. Principles of Biomedical Ethics (2001). Fifth Edition, Oxford University Press, UK

Dalton., D.D. Daily, C.M. Certo, S.T. & Roengprity, R. (2003), Meta-analysis of financial performance and quality. Fusion or confusion? Academy of management journal, 46: 13-16.

Dalton, D.D. Daily, C.M. Johnson, J.L. & Ellstrand , A.E. (1999). Number of directions and financial performance. A meta-analysis, Academy of management journal, 42, 674-686.

De Melo-Martin, I 2008, " Ethics, Embryos, and Eggs: The need for more than Epistemic values." American journal of bioethic occupation example, 8 (12): 38-40.

Douglas, H.E. 2003, " The moral responsibility of scientists (Tensions between autonomy and responsibility). " American philosophical Quarterly 40 (1): 59-68.

Gert, B. (1988) . Morality: A New Justification Of The Moral Rules, New York: Oxford University Press, USA

Rhodes, R. 2012, the Making Of The Atomic Bomb, 25. Anniversary edition. New York, Simon & Schuster.

Ziman, J. 2001. " Getting scientists to think about what they are doing." science and engineering ethics. 7 (2): 165-176.

IV

AI fifth stage development

Can non-manual driving public transport tools bring global economic growth

Why MTR underground train transportation needs to know passenger behaviour

Understanding individual passenger behaviour is essential for the design MTR transportation, because who can choose to catch bus, taxi, tram, train ferry etc. different kinds of public transportation tools. Individual traveler who decides to catch which kinds of public transportation tools, it depends on whether the public transportation tool can provide real time travel information, liking link travel time schedule. So, MTR underground train needs to understand where it has terminal to give convenience to the local living areas of time travelers to choose to catch MTR easily. Although, MTR ticket fare is one factor to influence any passengers choice. But, those other factors can also influence them to choice. e.g. MTR any terminal location of convenience, short time travelling, none crowding in busy (peak) time, MTR platform waiting arrival time, none sudden MTR engineering machines broken accident events occurrence frequently etc. different factors, any one of these factors which can influence passengers who choose to catch MTR or other kinds of transportation tools.

Why route choice can influence passenger behavioural choice
Usually, the busy time passengers will regard the route choice as a coordination problem to influence them to choose to catch which kinds of transportation tools. The route choice is as an opportunity costs to influence any busy time passengers to decide to choose to catch which kind of transportation tool which is the best right choice in the right time among of them. In the short time, for example, it seems any busy time passengers will choose to catch bus to substitute MTR underground train transportation tool, due to who feels the bus can arrive any destinations to compare other kinds of transportation tools in the most short time. However even if the MTR can either charge cheaper ticket fare to sell full day or charge discount ticket fare to sell in the busy (peak) time to compare to bus fare. It is possible that the busy time passengers will still choose to catch bus, if between the bus terminal and the another bus terminal that distance is the shorter time route to spend time to arrive destination to compare between the MTR terminal to the another MTR terminal arrival time . Also, although the busy time passengers will feel to enounter traffic jam to influence sitting or waiting bus time to be longer time in possible and who also feel MTR can avoid traffic jam problem. However, usually any busy (peak) time passengers will feel the chance of traffic jam occurrence will be less. So, the short bus route choice is more potential factor to influence the busy (peak) time passengers still to choose bus to catch.

However, if anyone wants to investigate results of day-to-day route choice which can be transferred to more realistic environment. It is necessary to explore individual behaviour in an interactive experimental set up to ensure busy (peak) time passenger transportation behavioural choice. For example, a passenger has a choice between a main road (M) and a side road (S) for travelling from (A) to (B). (M) is faster if (M) and (S) are chose by the same number of passengers. So, this method can be researched whether MTR terminal station is located at the main road (M) or the side road (S) where is more suitable to accept to passengers generally.

Why trip time reliability and
crowding factors can influence
MTR passenger choice.

Other problem is MTR busy (peak) time's crowding in public transportation occurrence of MTR underground train transportation tool is becoming a growth to concern as MTR demand growth at a busy (peak) time. To capture the MTR passengers benefits with reduced crowding from improved MTR public transport service and image. It is necessary a identify the relevant dimensions of crowding that are meaningful measures of what crowding means to MTR passengers. Two main influences on MTR model choice that are growing in relevance are trip time reliability and crowding. It represents a benefit-cost framework. In fact, MTR passengers can be willing to pay more expensive ticket fare, it MTR can avoid crowding and short and the accurate arrival trip time between terminals is reliable to occur. How to measure of MTR crowding, e.g. weighting the gap between the busy time, the standard (i.e. objective) and the perceived (i.e. subjective) metrics. We are not in a position to definitely map the two dimensions, which is a crucial requirement for translating objective improvements into equivalent subjective gains that then can be applied, willingness to pay estimates MTR ticket fares to obtain the additional MTR passenger benefits of MTR public transportation investment to any terminal stations. Because MTR crowding has a negative impact on passengers in terms of psychological on emotional distress. MTR passengers are willing to stand for up to 20 minutes of the service is fast and reliable. However crowding outweighed these benefits from a MTR passenger's perpective, experienced crowding leads a increased dissatisfaction. e.g. stress and less privacy during who needs to stand up in MTR. Due to there are no enough places to supply to them to stand up in MTR. If the

MTR trip time was longer time between the passenger's terminals, who will feel more dissatisfaction and it will cause who feels whether who ought need to choose to catch other transportation tools to substitute MTR next time. e.g. bus, train, tram, ferry, taxi etc. So, from an operator's perspective, the MTR service frequency or MTR size is significantly influenced by the level of ridership, which sends a signal to respond if the monitored crowding level exceeds the benchmark standard in the busy time. e.g. in the morning time or at the night time, the students or employment people who need to go to schools or offices (working places). The locations of different places between MTR terminals and crowding are regarded as a key service attribute for MTR pubic transportation along with other factors, such as travelling time and reliability, e.g. service quality, none engineering machines are broken to cause MTR stops suddenly.

Given the increasing importance of crowding on both the disutility to existing MTR public transportation users and the influence to it. MTR passenger can choose to use either the MTR public public transportation or other public transportation. It is timely to review the MTR current measures of crowding defined by transportation authorities. MTR operators ought evaluate whether they apporpriately reflect MTR each traveler experiences and perceptions of crowding in busy (peak) time. I suggest that MTR needs to buy other underground trains to supply to the busy (peak) time passengers to let them have enough seats to sit down, so who do not need to stand up in any MTR underground trains when they catch MTR underground trains in busy time. It aims to let who are willingness to pay the estimation of reasonable ticket fares to compare the other kinds of transportation tools in the busy (peak) time.

What is the crowding difference
between train and MTR underground train.
In fact, crowding won't be happened to brother these transportation tools easily in the busy time and non busy time both. e.g. bus, taxi, train, tram, ferry. Because passengers can not choose to stand up in

these transportation tools easily, due to these transportation tools have no enough areas (spaces) to let them to stand up easily . So, the crowding will be avoided to occur in these tranportation tools usually. Otherwise, MTR will have many passengers who can choose to stand up because MTR design of length is very long and it has enough areas (places) to let passengers to choose to stand up, even there have none any seats are provided to let them to sit down. So, MTR passengers will feel more dissatisfaction and crowding easily, especial in any peak (busy) time every day.

Comparing to bus, much more diverse crowding measures are defined in the passenger rail industry. For passenger, different specifications for measuring crowding are found across countries and even within a country. For example, rail crowding measures in the UK, the passengers in excess of capacity is crowding measure that applies to all London and South east operators weekday train services at a London terminus during the morning peak from 0700 to 09: 59 , and those departing during the afternoon peak from 16:00 to 18:59 (office of rail regulation 2011 year). The overall PIXC figure is considered the planned standard class capacity of each train service as well as the actual number of standard class passengers on the service at the critical point. i.e. the location on a trains of standard class passengers that surpass the planned capacity as the difference between the number of actual passengers and the capacity of the train divided by the number of passenger is within the capacity . So, it seems train and MTR underground public transportaton tools had been encountering the crowding problems in peak time, the difference in train passengers need to wait next train or more train arrival is who doesn't plan to enter the train, when who discovers the current train has no seats to provide to them to sit down in whose trip. Otherwise, MTR passengers can choose either to stand up within the large areas (places) if who discovered there are no any seats to provide to them to sit down or who can wait the next MTR arrival in order to who can sit down. It seems MTR transportation tool crowding environment includes in waiting platform and inside of the MTR underground train. Otherwise, train transportation tool crowding environment only includes the waiting platform and the passengers will not have crowding feeling inside of the train, due to none of passengers choose to stand up inside any trains because any train inside has no enough places to let them to stand up.

How MTR can attract many passengers.

On the commuter departure time choice of any reference point researching hand, the departure time decisions of communters are of fundamental importance of peak period MTR traffic congestion. However, whether on the demand side, MTR underground train congestion relief measures, such as MTR ticket fare to every terminal station needs to be charged cheaper fare or discount fare in the peak (busy) time every day. To aim to attract many passengers to choose to catch MTR Underground train public transportation tools, substitute to choose other public transportation tools in the peak time.

Over the past decades, there have been very active research efforts in the departure time problem, both in econometric modeling and dynamic user equilibrium fields. Although, these works provide valuable insights into dynamic commuter decision making, they do not identify the commuters' response to gains and losses related to whole actual arrival time to reference points who may have relative. The appliability of the reference point hypothesis of prospect theory to the commuter's departure time decision making to obtain a better understanding of how departure time choice in MTR platform during their waiting underground train arrival time. However, every MTR underground train actual arrival time and deviation variables related to reference points (gains and losses) are the key factors in the departure time choice model. How the MTR underground train of every communter's daily departure time decision can be modelled when the reference point hypothesis of prospect theory. The MTR underground train's schedule delay is defined as the difference between the preferred arrival time (PAT) and the actual arrival time (AT) for a given MTR communter. In a daily MTR commute, a commuter in the indifference band actual arrival time is an essential feature of MTR schedule study. Two reference points are the earliest acceptable arrival time and the work starting time for a given MTR platform waiting passengers. In psychological view point, prospect theory proposes that the displeasure of a loss is perceived or greater than the pleasure of a gain of the same attitude and therefore, the value function is stronger for losses than gains.

To conclude, it seems that if MTR waiting passengers need not spend long time to wait underground train arrival in platform and it can provide seats to let them to sit down in the busy (peak) crowding time. It will make them to feel pleasure, even the MTR ticket fare is not fair and reasonable to charge higher fare to compare other kinds of public transportation tools fares. So the peak waiting time factor can influence the passengers to choose other kind of transportation tools to catch easily. Moreover, MTR's two reference points are the earliest role. Similarly a loss is observed when the MTR platform waiting commuter experiences or actual arrival time which is beyond that the MTR schedule time. Due to that a MTR waiting commuter is as an early side arrival of whose actual arrival time is earlier than whose preferred arrival time.

Reference

Bailey, L., Mokhtarian, P.L. Little, A. (2008). The broader Connection Between Public Transportation, Energy Conservation And Greenhouse Gas Reduction, Report Prepared As Part Of TCRP Project J-11/Tasks Transit Cooperative Research Program, Transportation Research Board Submitted To American Public Transportation Association in http://www.apta.com/research/into/online/land_use.cfmi, accessed 17 April 2008.

The UK Standing Advisory Committee On Trunk Road Assessment (SACTRA) (1999). Transport And The Economy (Report To UK DETR). Retrieved From: http://webarchive.nationalarchives.gov.uk/20050301192906 ; http://dft.gov.uk/stellent/groups/dft-econappr/documents/pdf/dft_econappr_pdf_022512.pdf

Wikipedia Contributors (2008). Arterial Roads In Wikipedia, The Free Encyclopeda, http://en.wikipedia.org/w/index.php?title=Arterial_road&oldid=212832640(accessed May30,2008).

What the psychological need differences between rail and bus passengers

- Reasons we need to improve public bus transport tool service quality

The ways that we need to improve public transport, e.g. bus transport service, we try our best to ask these questions: During periods of stress on the bus, like weather conditions or maintenance failure that slows the bus service system? How to improve mass transit on bus service frequency, when looking at ways to improve public bus service transport , riders want frequency? Interestingly, speed is not as much of an issue, if they are waiting downtown in the rain, or on some suburban backstreet, riders want to know that a bus will arrive soon, preferably in less than 15 minutes. Therefore, the wait becomes part of the transportation cycle. Even, if the bus is lightning fast, in the mind of the rider, the trip begins right when they arrive at the bus station, and start waiting for the bus to pick them up.

`

`What does efficient bus ticketing system mean? It is big part of how to improve bus transportation efficiency is improving transit ticketing system, because ticketing systems have to be quick and practical to allow for prompt loading and unloading of passengers. So, inefficient ticketing systems also slow down bus frequency, as drivers need to wait for everyone to tap before they can drive away to the next stop.

How to let passengers feel comfortable? Riders want comfortable buses that can seat as many people as possible. Face-to-face seating is not appealing and being knee-to-knee in a confined space creates awkward moments between strangers. However, comfort also extends beyond the buses' seating arrangements. A smooth riding, quiet bus plays a significant role in reducing the overall stress of a public transit experience. Among the consistent feedback from riders of fuel cell electric buses is a surprised delight about how quiet the buses are when in motion.

On reduce greenhouse gases environment prote3ctoin aspect, exhaust spewing buses are on ongoing concern. One of the significant factors that commuters consider when deciding to take public transit is the environment impact of their alternative transport method. And although a diesel bus packed with 40 people may be less environmentally damaging than 40 separate diesel cars, it will still have negative impacts on both local air quality and the overall climate situation , when given the choice, we've found nearly all riders prefer " zero-emission buses" to conventional diesel buses nowadays.

IN fact, we are always thinking of ways to improve public transportation by dev4eloping new clean fuel technologies. Fuel cell electric buses resolve some of the above issues for both transit bus operators, bus performance is continually being proven and improved over millions of miles of operation in environments ranging from mountain villages to desert communities to busy cities. Hence, the first step to creating better public transit networks is becoming aware of the available options. Many communities are taking measures to improve public transport by implementing innovative sustainable transport solutions that have profound impacts on the live ability of their communities.

So, I shall recommend these ways to improve public transport methods to bus service as below:

Firstly, making interchanging easy for public transport has most efficient public transport service improvement aim at linking areas that are outside a city to the city center., doing this is beneficial in two ways. It helps people who should not at the city center , but needed to pass through because the outlying areas are not connected together to keep off and hence reduce congestion at the center. Also, connecting the outlying areas provide a backup for the public transport system in case of a problem which often happen. Secondly, minimize the number of stops/ stations, stops and stations improve the efficiency of public transport , but there should be a balance between enabling accessibility with more steps or stations and reducing the costs of operation by increasing transit need of ensure trips are covered in time. Therefore, core should be taken to ensure that stops and stations are located on streets to balance accessibility by commuters on one hand and reduces operating cost on the other hand.

Thirdly, lessen traffic congestion by deploying a number measures. Reducing traffic congestion at city streets could be done, implementing a number of strategies, such as providing lanes dedicated specially for the use of public transport, deploying strict regulations , such as queue bypasses or queue jumps. Another means of reducing traffic congestion is by providing feeds and data from public transport systems, freely to commuters to educate and help them avoid areas of traffic congestion and finally, giving priority to public and trams operating efficiency, increasing the travel time of these engineering mechanism whereby a traffic signal turns green at the light of a public transport at an intersection. All of above these improvements may be future public transport bus passengers service improvement need, if any bus companies hope to increase their bus passengers number absolutely.

● What rail passengers really want rail innovation improvement

Public transport systems, such as rail provides benefits including less traffic congestion, less pollution, safe travels, lower expenditures , less effort and better predictability in comparison to road transport. In fact, bus and train riders experience the most negative emotions in comparison with other transport modes, such as private cars , walking and cycling. Hence, technology has the potential to bring about the changes, needed to increase efficiency of rail transport, e.g. cost-effective ways to improve the quality of public transport and increase ridership may involve comfort and convenience improvement, or technology has the potential to provide more up-to-date information and customized service to train passengers and therefore improve the rail journey experience . On the overall, passenger journey , e.g. the importance of automated traveller information systems, and electronic fare payment collection systems can bring rail passengers look for this information in different interfaces from localized displays installed on platforms to smartphone applications.

Moreover, technology can also improve fare collection and management which of made manually can be prone to error, and time consuming , unified cards, smartphones can make it easier for rail passengers to obtain ticket, with the potential to increase the user satisfaction with the rail system. Because rail passengers demand not only pre-trip information for planning their travels, but also information during journeys, such as punctuality, connections and platform allocation. One extensive review indicates that accurate communication, for example, giving effective way finding information, can optimize passengers' experience with public transport.

Also, technology can facilitate the process of finding free seats on trains, which is a current demand from rail passengers and the cause of stress during the boarding process. IN fact, many rail passengers have specific preferences regarding seats and would appreciate having control of where to sit. So, navigation and way finding information can be delivered directly to passengers to inform where they could stand aiming to board less busy carriages, for example, choosing to travel on a less crowded train, or spreading themselves out on the platform before boarding in respond to crowding information, e.g. smartphones are frequently used by passengers of public transport and can make waiting times seem shorter. Furthermore specific system features designed for train passengers have the potential to improve the journey experience of the travelling public.

What ferry passengers service improvement need

- How can ferry service be improved affordable, reliable, convenient, flexible and clean will get drivers out of their cars ad onto environmentally responsible to passenger ferries?

Ferry transportation provides an environmentally friendly commuting alternative to the congested roadways in many of countries , so ferry transport service needs to meet long term air quality goals, it is critical to move beyond traditional technologies to zero-and near zero emissions technology. Clearly putting a transit system in operation that demonstrates emission control technology and the development of zero-emissions, ferries will help achieve air quality goals to our societies, for example., new shipping rout4es are needed to increase in order to satisfy ferry passengers different rapid ferry journey short distance need, when they need to choose one kind of public transport service either bus or rail or ferry transport service among of them.

None ferry accident occurrence, ferry service needs to let passengers to feel it is the safest sea pubic transit, expanded recreational service is also needs, particularly on weekends when bridge , corridor traffic congestion is becoming an increasing problem. Ferry service needs have uniquely provided flexible, vital transportation supports in response to a natural or man-made disaster that shuts down bridges and roads, fuel –cell technology is needed , that will lead to zero-emissions ferries, e.g. on-board emissions monitoring is far less polluting than previously through, e.g. 149 passenger boats are designed to travel 25 knots or less , and 300-350 passenger vessels designed for speeds up to 30-35 knots.

This emissions standard will perform specifications and the cost of this technology is accounted for in the ferry company vessel capital budget ,e.g. vessel design capabilities to accommodate existing and new docking configurations . This maximizes fast ferry passenger loading, including bicycles, carriages and wheelchairs. Hence, future global ferry service needs have these positive influence to our societies: Need for flexibility, desire to help the environment, need for time saving, which includes the importance of reliability, sensitivity to personal travel experience, such as a need for personal space or quiet feeling ferry seat any time, insensitivity to transport cost, e.g. the ferry ticket price is cheaper than rail or bus fares sensitivity to stress.

However, ferry service is different unlike rail, bus because expanded ferry service can be launched quickly at low initial cost and with great flexibility. Unlike buses, ferries are not hindered by traffic congestion on roads and highways or in tunnels. So, ferry service can be safely expanded to bring new service to new places and add more service to existing routes more easily than bus and rail public transport both, e.g. expanded ferry transport service can operate safety and provide with a robust, flexible and effective emergency response capability if the region is hit with a natural or man-made event that disables roads, other transit, bridges , before any.

Hence, ferry companies need to decide to improve their ferry transport service, they need to answer these questions: Is the new shipping route a good transportation investment? Does the new shipping route have fatal environmental negative impact? Does it offer a transit option that can be initiated in a timely and cost-effective manner? Can it provide ferry transport service that is reliable, safe and fully accessible after the ferry recovery would be unreasonably high charge to ferry selection is decided to implement to increase?

Also, ferry safety is needed to consider because it can influence any ferry passenger choice, when the ferry is moving on the sea, when the passenger is sitting on the boat. The ferry safety issue may include: Ensuring that access to all ferry operational areas, including, machinery spaces, pilothouse and gear lockers, remain locked at all times and accessible only to authorized crew, posting night watch security guards at terminals, conducting diligent onboard inspection for unattended passenger bags, briefcases and packages after each run, before the next boat load is allowed to board, creating coded signals and response to report suspicious activity, requiring positive identification before allowing any contractors, vendors or others access to ferries, providing additional security training to crew, developing a security plan to account for potential threats, outlining preventive measures and detailing an action plan in the event of a threat or actual emergency.

Future Human Transport Need Change

How future our transport need change? What factors influence our future transport need change? In general, these factors may influence our transportation need change. They may include fuel cost, the labor market for commercial drivers, demand for frieight , customer loyalty , vehicle capacity, government regulation, geographical events, the public transport tool reputation to passegners as a merchant. However, the factors that influence the development of transport system in an area? They may include as below:

Environment at the local scale existing hydrographical and geomorphological characteristics are string, factors in transport development, particularly in terms of the technical challenges (bridge, gradients,) they present to construct, other factors may include historical, technological, political and economic factors. All of these factors may influence our future transport system how develops. For raiway development influential factors, they may include: Geograohical factors, e.g. the North Indian plain with its level land, high density of population and rich agriculture presents the most favourable conditions for the development of railways in India. However, the presence of large number of rivers makes it necessary to construct bridges which involve heavy expenditure to Indian Government publich transport expenditure.

How transport has changed from past to present?

There has been a remarkable development in modern

transportation. The stream engine and then the stream trains have emerged and spread at this time and in abundance until the discovery of natural gas and oil was an evolution of transportation. Thus, the sedams and vehicles began to run in oil, until present battery changes energy vehicle need, even future non-manual driving artificial intelligent driving vehicle need. These new transport technology may influence our future public transportation from gas energy to battery changed energy, even non-manual driving vehicles need to our daily transport need.

So, our future purpose of public transport need is the unique purpose to oversome space, which is shaped by a variety of human and physical constraints, such as distance, time. These both is our future main public transport need main purpose factors, short distance and reducing journey time, they influence that why we need to choose to catch any kinds of public transportation tool to replace purchase private cars to drive transport tool choice. So, future any kinds of public transport tools, they need to consider above both main factors , how to attract passengers to choose to catch themselves public transport tools choice in this competitive public transport tools market.

On the other hand, the economic importance of transportation development can be defined as improving the welfare of a society, through appropriate social, political and economic conditions , such as US Government spent too much money to assist MTR (MAss transport railway firm) to develop underground thrain transport. Its aim to let many passegner can reduce journey time and reduce distance between destinations, it also hopes US citizen passengers can pay cheap transport fare to buy ticket to catch underground transport train for many families their transport expenditure in social transport welfare view.

However, US Government neds to solve those challenges, before it implements to develop rapid underground railway , e.g. lack of knowledge of geographical fwatures, lack of manpower necessary to operate the rapid underground railway construction work, lack of construction materials within the US itself. For Brazil rail network transportation development example, the factors influence the use of rail network for transportion is highly restricted in Brazil. Thus, the development of roadways and waterways is the main modes of transportation that caould be used in Brazil given its topography and drainage benefit to society . So, brazil can develop rail network for transportation development in success.

So, transportation system is important in the development of any nation, because transportation plays important role in rapid economic growth of a nation. Thrapsortation increases the quality and variety of consumer goods, thereby stimulating the demand and development of trade and economy of the nation. Moreover, transport provides various employment opportunities and boosts up the economy of the country.

Also, any transport tools need to improve themselves transport service in order to attract passengers to choose their public transport service more easily. They may attempt to sign up for an autonomous vehicle pilot program, free phone enquiey concerns whether the passegner can catch which bus bumber to go to the destination, hou much bus fare, how long journey time, when the bus will arrive teh bus stops or leave the bus stop etc. bus service questions, before any one passenger prepares to choose to catch bus (free bus go phone call enquiry), free download a public transport tool transit app. even water taxi tranport tool innovation can replace ferry public transport tool, it can let passengers have more fun an enjoyable catching feeling. So, water taxi tranport tool is one kind of future new transport tool change to replace ferry , it can influence ferry passengers to choose water taxi public transport tool to replace ferry. Although, its fare may be more expsnse to compare ferry, but it can reduce jounrey time and distance between both water stations, when ferry can not arrive the other destinations, but water taxi can arrive any one water station destination. It can bring convenient to future any one ferry passengers. So, water taxi may be developed to some countries, e.g. New Zealand , Auckland city, US , Washington and New York cities they had developed water taxi public transport tools to let ferry passengers have one kind new water public transport choice.

However, instead of new transport innovation improvement to water transport service public transport with input from the public on bus transport service aspect, bus frequency improvement, it means when booking at ways to improve, bus frequency from long times to less times, efficient bus ticketing system, a big part of how to improve tranportation efficiency is improving transit ticketing system.

In fact, my future transport system may still include these five types, modes of transport are: railway, roadways, airways, waterways and piplelines. Also, among different includes of transport, railways are the different modes of transport, railways are the cheapest. Trains cover the distance in less time and comparatively, the fare is also less to other modes of transporation. Therefore, railways is the cheapest mode of transportation to compare ferry, water taxi , sea transport, bus, taxi, road system.

On conclusion, transport price is not the main factor to attract passegners to choose to catch. The importance to have a good public transport system in place. It may be one main factor to help the kind of public transport tool to attract passengers to choose to catch, because a good transport links can widen people's job search area and help them find employment. It can also reduce commuting times and reduce the cost of living, and high skilled workers are more likely to travel across longer distances to work, especially if they are following good job opportunities. So, future any one kind of public transportation tool service provider ought consider how to satisfy working people working time need to shorten journey time to any working places or student learning time need to shorten jounrey times to any schools as well as let they feel comfortable to sit on comfortable chairs or provide free internet service to themselves mobiles , laptops, when they are sitting down or standing up in the kind of public transport . It is the important factor to influence any kind of public transport service in success.

Future Non-Manual driving vehicle How
Influences Public Transport Tool Passenger Need

Nowadays, artifical intelligent (non-manual) driving vehicles are invented, it may be accepted to any countries families to feel comfortable to drive on roads, because any people choose to buy any kinds cars, when any people choose to buy kinds of non-manual (artificial intelligent) vehicles, they do not need to use their hands to drive cars, because artificial intelligent (robotic auto control wheels, it means that robots can help human (drivers) to control wheel to drive to avoid any cars crash occurrence on the roads more easily.

If one day, non-manual driving robotic control whoole vehicles are invented in successful, whether it will persuade many different conuntries families choose to buy non-manual (robotic auto control wheel) vehicles, then it will cause bus, tram, train, underground train, road transport need will be influenced to reduce or even if non-manula boats are invented, whether it will cause ferry sea transport needs will b influenced to reduce. Hence, future non-manual driving vehicles or bats invention whether they will influence public transport tool of road and sea transport passengers number reduces. It is one interesting question. I shall attempt to discuss as below:

In fact, non-manual vehicles are very attraction, to excite any person chooses to buy to drive, because people do not need often touch wheels and touch foots button to control cars to move often forever, when robotic can be invented to help human to control car wheel and foot button, any person only needs to sit on his/her car, then the car can move rapidly, because any drivers is lazy, he/she hopes machine can help her/him to drive car on the road safely. So, he/she can read book or listen music or eatch mobile movie to enjoy his/her entertainment when he/she is sitting on his/her car.He/she will feel more comfortable and enjoyable when robotic can help him/her to drive car. So, robotic (non -manual driving vehicle) can encourage people to choose to buy cars because any drivers won't need to drive cars, robotic can help drivers them to drive on the road easily, when global any one family can own one robotic auto control (non-manual driving) car at least, it may influence these owning non-manula diriving vehicle owners do not feel need to pay any fares to buy road public transport tools of bus ticket, train ticket, underground train ticket , tram ticket to go to anywhere. So, it seems that robotic (non-manual driving) vehicles may influence future any road transport passengers number reduces , because traditional catching any kinds of road public transport tool passengers will be influenced to choose to sit themselves auto (non-manual) driving cars to go to offices to work, parents do not need to follow their sone/daughters to sit on themselves non-manual auto driving cars to go to schools, because their sons/daughters can sit on themselves non-manual driving cars to go to schools more easily. In holidays, they can sit on themselves non-manual driving cars to go to cinemas, music halls, breachs, theaters, shopping centers, gardens different entertainment places to enjoy their any leisure safely because robotic can help them to drive their cars on roads safely.

So, it means that robotic auto control driving cars can influence global every family to feel that they do not need to catch any kinds of public transport tools, e.g. bus, train, tram, taxi underground train to go to anywhere because robotic auto driving cars can help any one, he/she does not know how to drive car to go to anywhere safely. So, future any one won't need to learn driving car skill, when he/she likes to buy one auto driving car. So, in passenger public transport need view, non-manual driving cars will influence them to feel any kinds of road public transport tools can help them to go to anywhere conveniently, because themselves non-manual driving vehicles can help them to drive cars to go to anywhere conveniently. They only need to tell robotic that where they want to go, when they sit on their non-manual driving cars, then robotic knows whether where destination, they want to go, their cars will auto move on the road immediately. It is one exciting and enjoyable ourney when the driver does not need to drive his/her car on the road. So, it seems that robotic (non-manual driving) vehicles invention may bring negative influence to any kinds of public transport tools service needs to passengers , when passengers had owned one non-manual driving car at least.

Why and how non-manual driving car owners need

raise public transport quality on travel time and fare

aspects

● How non human driving behavior can be influence by non-manual driving cars

In fact, impact of automated vehicless on travel mode preference, it can bring both trip purposes and distances aim raising need to any kinds of public transport service passegners. Because of technology penetration in the transportation system, the automated vehicle is set to be a future mode of transport, it may bring negative impact to future any kinds of public transport passengers needs, in special on the potential impact of these non-manual driving automated vehicles on travel behaior negative impact to public transport passenger behavior. Automated vehicles will influence future public transportation passengers feel it can bring more short time travel distances and short trip purposes more benefit than any kinds of public transport choices, e.g. bus, taxi, ferry, train, tram, underground tram etc. road and sea public transport tools, e.g. ferry, water taxi. It means that when future any passenger feels above these any one kind of public transport tool needs to spend longer travel time on journey distance and trip to compare future automated vehicles, then they will choose to sit on automated vehicles in preference, due to automated vehicles can help global any one person needs to go to anywhere rapidly.

So, automated vehicles may replace general traditional public transport tools in possible, when they are popular accepted in societies. On the other, instead of shortening journey travel distance time, (travel time) aspect, public transport fare, travel cost will be another influential factor to influence future public transport tool passengers to choose automated vehicles to replace to catch any kinds of public transport tools.

In fact, conventional cars and public transport s are perceivd as being the least attractive alternative in relation to in-vehicle travel time on short and long distance communting trips. So , future automated vehicle drivers (non -human driving) behaviors will be likely changed to prefer this mode for long distance leisure trips rather than short distance commuting trips by automated vehicles.

In fact, advanced technologies have revolutionized many aspects of human life, include the automated vehicle transport system. Also, transport system is one of the essential development aspect to particular , such as non-manual driving automation , vehicle aims to make trips safer, faster , more efficient, automated vehicles passengers and drivers can feel enjoyable to do themselves leisure behavior , e.g. read books, listen, music, listen mobile, watch laptop movies when any one does not need to consider whether their cars are safe to be driven , even any one needs to drive the automated car, because robotic can help them to control how to automatic drive this car on the road safely.

Robotic will bring confidence to let them feel that themselves cars are moving safely on the roads . In recent years, the concept of automated driving has been introduced as on outstanding platform for the next generation of driving systems that is expected to improve safety, traffic flows efficiency, reducing traffic jams occurrence chance, avoiding traffic accidents occurrence chance, e.g. avoid to crash any one person when he/she is walking across road or crach any car is moving on the road easily, capacity, accessibility , and reducing congestion through the application of some technologies , such as vehicle to vehicle and vehicle to infrastructure communication.

So, future automated vechicles can have good driving facility systems to be installed in their cars, in order to raise safety, rapid driving speed level to let any one to feel , when they are sitting in their automated cars, e.g. using cameras, sensors, global positioning system adaptive cruise control, light detection and ranging, and advanced driver assistance system, automated vehicles can steer the vehicle and drive it automatically when passengers delegate control to a computer. Absolutely, ny replacing the driver role with an automated driving system , future one automated vehicle is able to totally free up passengers under automation levels.

So, unless future any kinds of public transport tools may apply automated robotic automated driven system replace the bus driver, taxi driver, train driver, tram driver, underground train driver to raise automated driving system service improvement level to let any one passengers to feel. Otherwise, when automated vehicles are popular to be accepted to buy in any one country in global. Then, global public tansport tool passegners number may be influenced to reduce when global any one family owns at least one automated vehicle at themselves homes .

In other words, automated vehicles can bring thes benefits to let global any one household family feels, future automated vehicles users , they can mostly behave like passengers inside the vehicle, which implies that they will be able to multitask and productive by allocating the travel time to do other activities, e.g. reading, eating, working, drinking, watching movies, listening musics, even sleeping. So, automated vechicles will motivate humans to change non-humanly driven behaviors from conventional humanly driven behavior. This non-humanly driven behavior may be one main factor to influence or encourage future any one kind of public transport passenger won't choose to pay fare to buy ticket to catch any one kind of public transport tool again, because non-manual driven behavior may hel many lazy people do not need to consdierate how to learn to drive cars skills to prepare pass any road test in order to earn the driving licnece to permit to drive cars forever. When automated vehiclesa re popular to be accepted to replace manual-driven cars in societies.

Hence, automated vehicles could potentially change the traditional human driven vehicle market to cause their manual driven cars sale buyers number reduces, when the automated vehicle buyers number increases, also they can chance globa public transport passengers behaviors to reduce to pay fares to catch any kinds of public transport tools when automated vechicles users may sit on themselves automated vehicles to go to anywhere in short time rapidly and safely in any countries.

On conclusion, future global public transport service competition is serious, because instead of global passengers had began to compare whether which kinds of public transport fares are cheaper, more safe, shortening journey time between leaving place and destination, more comfortable feeling, e.g. clean and comfortable chairs , mre free internet service facilities in order to make any one kind of catching public transport tool choice in preference. On the other hand, future automated vehicles number will increase when traditional manual driven car users begin to believe that automated vehicles can bring more safe , more comfortable, more fee- time using, more leisure satisfactory feeling, more than traditional manual driving cars. Then, when global any one household family had made choice to buy at least one automated vehice to replace themselves car(s) at home. When, they are habit to sit in themselves automated vehicles to go to anywhere, however, short or long trip . Consequently, global any one household family won't feel any kinds of public transport tools may bring personal economic saving cost, comfortable, enjoyable, free-time using benefit to compare themselves automated vehicles . It will cause global public transport tools passengers number will reduce , when many different kinds of home automatic vehicles are purchased to replace manual driving cars by global household automated vehicle users. So, in passegner transport tool choice psychological view, automatic vehicles will be possible to replace future public transport service tools. So, any public transport service providers can not neglect how to desing and improve their facilities , charge reasonable transport fare, provide more comfortable, and enjoyable sitting feeling , even applying automatic driving system to replace human drivers in order to attract passegners ' catching need choice more easily.

Prediction travel market changing method

What factors can influence travel behavioural consumption

Prediction travel behavioral consumption from psychology view and computer statistic view.

How to predict travel consumption? It is one question to any travel agents concern to use what methods which can predict how many numbers of travelers where who will choose to go to travel more accurately. I think that who can consider how to predict travel behavioral consumption from psychology view and computer science view both.

On the psychology view, It has evidence to support the relationship between self-identify threat and resistance to change travel behavior to any travelers, controlling for whose past travelling behavior, resistance to change if a psychological phenomenon of long standing interest in many applied branches of psychology. Past travelling behavior has been acknowledged as a predictor of future action. Such as travelling behavior that is experienced as successful is likely to be repeated and may lead to habitual patterns. Some psychologists differentiate habit between two concepts, such as goal oriented and automatic oriented both. Although repeated past travelling behavior is addition goal oriented and automatic oriented. Further non-deliberative nature of habit may make appeals to judge and to predict future individual traveler's behaviour accrately. However, repeated travelling behavior without a necessary constraint of goal orientation and automatic oriented both. So, it seems that psychological factor can influence any individual traveler why and how who choose to decide whose travelling behaviour.

On the computer statistic view, structural equation modeling is an extremely flexible linear-in-parameters multivariate statistical modeling technique. It has been used in modeling travel behavior and values since about 1980 year. It is a software method to handle a large number of variables, as well as unobserved variables specified as linear combinations (weighted averages) of the observed variable.

Whether climate change can influence travelling behaviours.

The flexibility of human travelling behavior is at least the result of one such mechanism, our ability to travel mentally in time and entertain potential future. Understanding of the impacts is holidays, particularly those involving travel. Using focus groups research to explores tourists' awareness of the impacts of travel own climate change, examines the extent to which climate change features in holiday travel decisions and identifies some of the barriers to the adoption of less carbon intensive tourism practices. The findings suggest many tourists don't consider climate change when planning their holidays. The failure of tourists to engage with the climate change to impact of holidays, combined with significant barriers to behavioral change, presents a considerable challenge in the tourism industry.

Tourism is a highly energy intensive industry and has only recently attracted attention as an important contributions to climate change through greenhouse gas emissions. It has been estimated that tourism contributes 5% of global carbon dioxide emissions. There have been a number of potential changes proposed for reducing the impact of air travel on climate change. These include technological changes, market based changes and behavioral changes. However, the role that climate change plays in the holiday and travel decisions of global tourists. How the global tourists of the impacts travel has on climate change to establish the extent to which climate change, considerations features in holiday travel decision making processes and to investigate the major barriers to global tourists adopting less carbon intensive travel practices. Whether tourists will aware the impacts that their holidays and travel have on climate changes.

When, it comes to understand indvidual traveler's behavioral change, wide range of conceptual theories have been developed, utilizing various social, psychological, subjective and objective variables in order to model travel consumption behavior. These theories of travel behavioral change operate at a number of different levels, including the individual level, the interpersonal level and community level. Whether pro-environmental behavior can be used to predict travel consumption behavior in a climate change. However, the question of what determines pro-environmental behavior in such a complex one that it can not be visualized through one single framework or diagram.

Despite the potentially high risk scenario for the tourism industry and the global environment, the tourism and climate change ought have close relationship. Whether what are the important factors and variables which can limit tourism? e.g. money, time, family problem, extreme hot or cold weather change, air ticket price, journey attraction etc. variable factors. Mention of holidays and travel were deliberately avoided in the recruitment process, so as not to create a connection factor to influence traveler's individual mind. However, the dismissal of alternative transportation modes can be conceived as either a structural barrier, in the sense that flying is perhaps the only realistic option to reach long-haul holiday destination, or a perceived behavioral control barriers in that an individual perceives flying as the only option open to whom. The transportation tool factor will be depend to extent on the distance to the destination. This can also be interpreted in a social perspective as an intention with the resources available where much international tourism is structured around flying. To increase the availability of different transportation modes, tourists could choose holiday destination closer to home.

Finally, also how to predict future travel behavioural consumption. I feel that travel agents need to predict whether any country's random daily variation of weather factor is also important to influence travel behaviour. e.g. in weather, temperature, rainfall adn snowfall with traffic accidents factors will have relationship to cause travel demand. Some scientists estimate suggest that when warmed temperatures and reduced snowfall are associated with a moderate decline in non-fatal accidents, they are also associated with a significant increase in fatal accidents. Thus increase in fatalities and temperature. Half of the estimated effect of temperature on fatalities is due to changes in the exposure to pedestrians, bicyclists and motorcyclists as temperature increase. So, if any countries have rainfall, snowfall and low temperature to cause traffic accidents, whether this accident occurrence will influence the travelers who liking climb snow hills, riding bicycle, running sports who will avoid to travel to these countries' bad weather after occurs. So, why I feel that this natural climate factor will also be one serious factor to influence travel behavioral consumption.

Future travel consumption behavior

Whether individual habitual behaviour can influence travelling behaviour : e.g. renting travel transportation tools

Whether habit can be intended to predict of future travel behavior to people are creatures of habits. Many of human's everyday goal-directed behaviors are performed in a habitual fashion, the transportation made and route one takes to work, one's choice of breakfast. Habits are formed when using the some behavior frequently and a similar consistency in a similar context for the some purpose whether the individual past travel consumption model will be caused a habit to whom. e.g. choosing whom travel agent to buy air ticket or traveling package; choosing the same or similar countries' destinations to go to travel ; choosing the business class or normal (general) class of quality airlines to catch planes. Does habitual rent traveling car tools use not lead to more resistance to change of travel mode? It has been argued that past behavior is the best predictor of future behavior to travel consumption. If individual traveler's past consumption behavior was always reasoned, then frequency of prior travel consumption behavior should only have an indirect link to the individual traveler's behavior. It seems that renting travel car tools to use is a habit example. So, a strong rent traveling car tools useful habit makes traveling mode choice. People with a strong renting of traveling car tools of habit should have low motivation to attend to gather any information about public transportation in their choice of travelling country for individual or family or friends members during their traveling journeys.

Even when persuasive communication changes the traveler whose attitudes and intention, in the case of individual traveler or family travelers with a strong renting travel car tools habit. It is difficult to change whose travel behaviors to choose to catch public transportation in whose any trips in any countries. However, understanding of travel behavior and the reasons for choosing one mode of transportation over another. The arguments for rent traveling car tools to use, including convenience, speed, comfort and individual freedom and well known. Increasingly, psychological factors include such as, perceptions, identity, social norms and habit are being used to understand travel mode choice. Whether how many travel consumers will choose to rent traveling car tools during their trips in any countries. It is difficult to estimate the numbers. As the average level of renting travel car tools of dependence or attitudes to certain travel package policies from travel agents. Instead different people must be treated in different ways because who are motivated in different ways and who are motivated by different travel package policies ways from travel agents.

In conclusion, the factors influence whose traveler's individual behavior either who chooses to rent traveling car tools or who chooses to catch public transportation when who individual goes to travel in alone trip or family trip. It include influence mode choice factors, such as social psychology factor and marketing on segmentation factor both to influence whose transportation choice of behavior in whose trip.

How to determine future travel behavior from past travel experience and perceptions of risk and safety for the benefits to travel consumers?

How to determine future travel behavior from past travel experience and perceptions of risk and safety for the benefits to travel consumers? Why does individual traveler avoid certain destination(s) is(are) as relevant to tourist decision making as why who chooses to travel to others. Perceptions of risk and safety and travel experience are likely to influence travel decisions. If travel agents had efforts to predict future travel behavior to guess whether travelers will feel where is(are) risk and unsafe to cause who does not choose to go to the country to travel. Then, the travel agents will avoid to choose to spend much time to design the different traveling package to attract their potential travel consumers to choose to travel. The reason is because in the case of individual traveler's tourism experience, the traveler whose past disappointment travel experience (psychological risk) will be a serious threat to the traveler's health or life (health, physical or terrorism risk). The past safety or unhealthy risk to the country(countries) will influence the traveler decides to choose not to go to the countries(country) to travel again in the future.

What is push and pull factors to influence any
traveler who chooses where is whose preferable travelling destination

How to predict individual traveler's behavioral intention of choosing a travel destination. Understanding why people travel and what factors influence their behavioral intention of choosing a travel destination is beneficial to tourism planning and marketing. In general, an individual's choice of a travel destination into two forces. The first force is the push factor that pushes an individual away from home and attempt to develop a general desire to go somewhere, without specifying where that may be. The other force is the pull factor that pull an individual toward in destination, due to a region-specific or perceived attractiveness of a destination. The respective push and pull factors illustrate that people travel because who are pushed by whose internal motives and pulled by external forced of a destination. However, the decision making process leading to the choice of a travel destination is a very complex process. For example, a Taiwanese traveler who might either choose new travel destination of Hong Kong or another old travel Asia destinations again or who also might choose any one of Western country, as a new travel destination. The travel agents can predict where who will have intention to choose to travel from whose past behavior and attitude, subjective and perceived behavioral control model.

The factors influence where is the traveler choice, include personal safety, scenic beauty, cultural interest, climate changing, transportation tools, friendliness of local people, price of trip, trip package service in hotels and restaurants, quality and variety of food and shopping facilities and services etc. needs. So, whose factors will influence where is the individual travel's choice. It seems every traveler whose choice of travel process, will include past behavior. e.g. travelling experience, travelling habit, then to choose the best seasoned travelling action to satisfy whose travel needs. This process is the individual traveler's psychological choice process, who must need time to gather information to compare concerning of different travel packages, destination scene, climate change, transportation tools available to the destination, air ticket price etc. these factors, then to judge where is the best right destination to travel in the right time.

Why expectation, motivation and attitude factor can influence travelling behaviour.

Social psychology is concerned with gaining insight into the psychological of socially relevant behaviors and the processes. For instance, on a global level bad influence to global warming, it influences some countries extreme cold or hot bad climate changing occurrence, then it ought influence some travelers' behavioral decision to change their mind to choose some countries to go to travel at the moment which do not occur extreme hot or cold climate (temperature). e.g. above than 40 degree in summer or below than 0 degree in winter. Due to the extreme climate changing environment in the countries, it will cause them to feel uncomfortable to play during their trips. So, the global warming causes to climate changing factor will influence the numbers of travel consumption to be reduced possibly. This is global climate changing environment factor influences to bad or uncomfortable social psychological feeling to global travelers' mind of traveling decision. What is individual traveler expectation, motivation and attitude? Tourism sector includes inbound (domestic) tourism and outbound (overseas) tourism both incomes to any countries. According to recent article, a tourist behavior model has been developed, called the expectation, motivation and attitude (EMA) model (Hsu et al., 2010).

This model focuses on the pre-visit stage of tourists by modeling the behavioral process by incorporating expectation, motivation and attitude. Travel motivation is considered as an essential component of the behavioral process, which has been increasing attention from the travel; industry. The economic approach defines "tourism" is an identifiable nationally important industry. It includes the component activities of transportation, accommodation, recreation, food and related service. So, tourism behavioral consumption is concerned the individual tourist's usual habituate of the industry which responds to whose needs, and of the impacts that both the tourist and the tourism industry have on the socio-cultural, economic and physical environment.

However, travel motivation means how to understand and predict factors that influence travel decision making. According to Backman and others (1995, p.15), motivation is conceptually viewed as " a state of need, a condition that services as a driving force to display different kind of behavior toward certain types of activities, developing preferences, arriving at some expected satisfactory outcome." So, motivation and expectancy which has close relationship to any tourist before who decided to do any tourism of behavior. Some economists confirmed motivation and expectancy which has relations, such as expectation of visiting an outbound destination has a direct effect on motivation to visit the destination; motivation has a direct effect on attitude toward visiting the destination; expectation of visiting the outbound destination has a direct affect on attitude toward visiting the destination and motivation has a mediating effect on the relationship in between expectation and attitude.

What methods can predict future travel behavioural consumption

How to use qualitative of travel behavioural method to predict future travel consumption.

I also suggest to use qualitative of travel behavioural method to predict future travel consumption. Methods such as focus groups interviews and participant observer techniques can be used with quantitative approaches on their own to fill the gaps left by quantitative techniques. These insights have contributed to the development of increasingly sophisticated models to forecast travel behavior and predict changes in behavior in response to change in the transportation system. First, survey methods restrict not only the question frame but the answer frame as well, anticipating the important issues and questions and the responses. However, these surveys methods are not well suited to exploratory areas of research where issues remain unidentified and the researched seek to answer the question "why?". Second, data collection methods using traditional travel diaries or telephone recruitment can under represent certain segments of the population, particularly the older persons with little education, minorities and the poor. Before the survey, focus group for example can be used to identify what socio-demographic variables to include in the survey, how best to structure the diary, even what incentives will be most effective in increasing the response rate. After the survey, focus, focus groups can be used to build explanations for the survey results to identify the "why" of the results as well as the implications. One Asia Pacific survey research result was made by tourism market investigation before. It indicated the travel in Asia Pacific market in the past, had often been undertaken in large groups through leisure package sold in bulk, or in large organized business groups, future travelers will be in smaller groups or alone, and for a much wider range of reasons. Significant new traveler segments, such as female business traveler. The small business traveler and the senior traveler, all of which have different aspirations and requirements from the travel experience.

Moreover, Asia tourism market will start to exist behaviors in the adoption of newer technologies, a giving the traveler new ways to manage the travel experience, creating new behaviors. This with provide new opportunities for travel providers. The use of mobile devices, smartphones, tablets etc. and social media are the obvious findings to become an integral part of the travel experience. Thus, quality method can attempt to predict Asia Pacific tourism market development in the future.

However, improving the predictive power of travel behavior models and to increase understanding travel behavior which lies in the use of panel data(repeated measures from the same individuals). Whereas, cross-sectional data only reveal inter-individual differences at one moment in time, panel data can reveal intra-individual changes over time. In effect, panel data are generally better suited to understand and predict (changes in) travel behavior. However, a substantial proportion was also observed to transition between very different activity/travel patterns over time, indicating that from one year to the next, many people renegotiated their activity/travel patterns.

How to apply advanced traveler information systems (ATIS) to predict future travelling behaviour.

Nowadays, information can impact on traveler behavior and network performance. For example, when steadily growing levels of vehicle ownership and vehicle miles traveled information has been identified as a potential strategy towards man aging travel demand, optimizing transportation networks and better utilizing available capacity. Toward, this goal to predict further tourist behavioral consumption. Many countries, government tourism development institutes has applied advanced traveler information systems (ATIS) which travel behavior models and high-fidelity network performance models made increasingly feasible through the rapid advances in computer power. Crucial components of this problem domain are the modeling of individual tourist drivers' response to travel information and the development accurate guidance of relevance to real would trip makers. So, this advanced traveler information systems (ATIS) can assist the tourist who like to rent travelling car tools to travel in any countries own free traveler information systems service conveniently. Also, this travel information system can be intended to assist travelers to make better travel choices. e.g. this system can improve the decision making of individual traveler rather than improvements of network performance overall. So, we need to understand how tourists make their travel plans. Also, understanding decision process that lead to booking of the trip is equally important, as it allows of a potential behavior.

How does online tourism sale channel can influence traveling consumption of behaviour.

Nowadays, internet is popular, it seems that booking air ticket behavior of using internet is predicted to influence overall tourism air tickets payment method. Tourism industry has grown in the previous several decades. Despite its global impact, questions related to better understanding of tourists and whose habits. Using online travel air ticket booking benefits include booking electronic air tickets can be made from entering any electronic travel agents websites in the short time and electronic travel ticket payers do not need leave home, who can pay visa card to pre booking any electronic travel ticket from online channel conveniently.

How to analyze activity based travel demand ? Nowadays, human are concerning the traffic congestion and air quality deterioration, the supply oriented focus of transportation planning has expanded to include how to manage travel demand within the available transportation supply. Consequently, there has been an increasing interest in travel demand management strategies, such as congestion pricing that attempts to change aggregate travel demand. The prediction aggregate level, long term travel demand to understanding disaggregate level (i.e. individual levels) behavioral responses to short term demand policies, such as ride sharing incentives, congestion pricing and employer based demand management schemes, alternate work schedules, telecommuting limitation of travel agent traditionally work nature shall influence oriented trip based travel modelling passenger travel demand indirectly.

Finally, online travel purchase will be popular to influence the number of travel behavioural consumption nowadays. Any travel package products can be sold from websites to attract travellers to choose to prebook air ticket for any trips conveniently. In the past ten years, the internet has become the predominant carrier of all types of information and transactions. Regarding travel decisions, internet has also become an important sales channels for the travel industry, because it is associated with comparably lower distribution and sales costs, but also because ir adapts to hign supply and demand dynamics in this industry. Consequently, the travel and tourism industry tries to increase the internet sale specific share of sales volumes. So, internet sale channel has changed travel consumption behavioural pattern and characteristics and travel experience. For example, Switzerland has one of the highest population-to-computer ratio in Europe. It is also one of the most highly internet penetrated countries in terms of use of the WWW on a day-to-day basis, with more than 75 percent of the population older than 14 years using the WWW daily (ICT, 2005).

The reason of booking online tourism may include: convenience, fast transaction, finding traveling package choice easily, more airline seats available. So, online booking tourism will influence the traditional tourism agents visiting of sales and air tickets and travelling package numbers to be decreased. Finally, the online booking tourism market shares will be expanded to more than traditional tourism agents visits sale market in the future one day. So, the travel agents who still use the traditional tourism visiting sale channel which ought raise whose features to compare to differ to online tourism sale channel if these traditional touriam agents want to keep competitive ability in tourism industry for long term.

Actively based patterns of urban population of travel behavioural prediction method.

Actively based patterns of urban population. It is a method of motivational framework means in which societal constraints and inherent individual motivations interact to shape activity participation patterns. It can be used to predict one city or urban the numbers of travel demand in the year. It has two elements: First, capability constraints refer to constraints are imposed by biological needs, such as eating and sleeping and/or resources, such as income, availability of cars etc. to undertake the urban or city's family activities in the year. Second, coupling constraints define where, when and the duration of planning activities that are to be pursued with other individuals. So, this method needs to gather information (data) to get the relationship between activities, travel and spending work time and space time to evaluate whether there are how many families who have real needs to spend time to go to travel in the year.

What is trip based versus activity based approaches?

What is trip based versus activity based approaches? The fundamental difference between the trip-based and activity based approaches is that the former approach directly focuses on trips without explicit recognition of the motivation or reason for the trips and travel. The activity based approach , on the other hand, views travel as a demand derived from the need to pursue travel activities. So, it is better understand the individual or family behavior basis for individual or family travelling decision regarding participation in travelling activities in certain places or cities or countries at given times and hence the resulting travel needs. This behavioral basis includes all the factors that influence the why, how, when and where of performed activities and resulting individuals and household, the cultural/social norms of the community and the travel surrounding environment.

Another difference between the two approaches is in the way travel is represented. The trip based approach represents travel as a

collection of trips. Each trip is considered as independent of other trips, without considering the inter-relationship in the choice attributes , such as time, destination and mode of different trips. As tours are chains of trips beginning and ending at a same location , say home or work. The tour based representation helps maintain the consistency across and capture the interdependency and consistency of the modeled choice attributed among the trips of the same tour.
In addition to the tour based representation of travel, the activity based approach focuses on sequences or patterns of activity participation and travel behavior, using the whole day or longer periods of time is the unit of analysis. Such as approach can address travel demand management issues through an examination of how people modify their activity participation, for example, will individuals substitute more out-of-home activities for in home activities in the evening of who arrived early form work due-to a work schedule change?

The major difference between trip based and the activity based approaches is in the way, the time dimension of activities and travel is considered. In the trip based approach, time is reduced to being simply a cost making a trip and a day's viewed as a combination, defined peak and off peak time periods. On the other hand, activity based approach views individuals' activity travel patterns are a result of their time use decisions with a continuous time domain. As individuals have 24 hours in a day or multiples of 24 hours for longer periods of time and decide how to use that travel among or allocate that time to activities and travel and with who, subject to their socio-demographic, transportation system and other and scheduling of trips. So, determining the impact of travel demand management policies on time use behavior is an important step to assessing the impact of such policies on individual travel behavior. The final major difference between this two approaches relates to the level of aggregation. In the trip based approach, most aspect of travel, e.g. number of trips etc. are analyzed at an aggregate level.

Consequently, trip based methods accommodate the effect of socio-demographic attributes of households and individuals in a very limited fashion, which limits the activity of the method to evaluate travel impacts of long term socio-demographic characteristics of the individuals who actually make the activity travel choices and the travel service characteristics of the surrounding environment. So, the activity based models are better equipped to forecast the longer term changes in travel demand in response composition and the travel environment of urban areas. Also, using activity based models, the impact of policies can be assessed by predicting individual level behavioral responses instead of employing trip based statistical averages that are aggregated over defined demographic segments.

Why senior age will be main travelling target.

In the past, Germany government had established tourism survey analysis to analyze survey data in order to arrive at reliable conclusions on future trends in travel behavior. To aim to find how demographic change will influence the tourism market and how the industry can adapt to those changes. The travel analysis provided data on tourism consumer behavior, including attitudes, motives and intentions. Since, 1970 year, it is based on a random sample, representative for the population in private households aged 14 years or older. Then, a continuous high scientific standard combined with a national and international users makes the travel analysis a useful tool and reliable source for tourism industry and policy decisions. It aimed to gather statistical data. e.g. on the age structure and on demographic trends, quantitative and qualitative analysis with time series data from the travel analysis. It shows e.g. not only the future volume , quite different from today's seniors, or how who will travel of family holidays will change, e.g. single parents of low, but grandparents of growing significance for tourism.

Demographic change is said to be one of the important drivers for new trends in consumer traveling change behavior in most European countries (e.g. Lind 2001). Because the growing number of senior citizens in the European Union and other industralised countries, such as the USA and Japan, looks to become one of the major marketing challenges for the tourism industry. United Nations statistics predict that the share of people being 60 age or older will grow dramatically in the coming future, and is expected to rise from 10 percent of the world population in 2000 year to more than 20 percent in 2050 year (United Nations Population Division, 2001). From its statistic, some data showed that travel propensity increased throughout life until the age of about 50 years of age and was then kept stable until very late in life 75 age. The most important results is that the travel propensity when getting older is not going down between 65 and 75 age of course, the overall development of this variable is influenced by a lot of other factors which are rsponsible for quite a variation over time. It is now possible to suggest that the general pattern of travel propensity is one of the key indicators for holiday life cycle travel behaviour, includes three stages. The growth stage tends to increase from early aduithood until 45 age old or when reaching some 80%. The next stage is stabilisation from the ages of around 50 age,until 75 age old, starting with a lower increase. Finally, the decrease stage is a slight decrease occurs once people reach the more advanced age of 75 age to 85 age old (Lohmann & Danielsson 2001).

So, it seems Germany government tourism prediction to future travellers' behaviour indicated these findings, such as on how future senior generations will travel, who had used survey data to examine the patterns of travel behaviour of a generation getting older and applied the findings to draw conclusions on the future. Also, it predicted that on the future of family trips, family semgmentation will be the travel behaviour patterns in the future. These findings together with the statistical data on demographic change allowed for a better understanding of the coming tends in family holidays. It's aim developed in consumer behaviour related to demographic change and predicted what will happen future of tourism one had to consider other influences and drivers as well, for example, trends on the supply side. e.g. low cost airlines or in travelling consumption behaviour in general whether how the past may provide a key to predict travel patterns of senior sitizens to the future.

Given the projected growth of the senior citizens market, designing specific marketing strategies to meet the prospective needs of elderly tourists will become increasingly important. It has been an implict assumption that it will be a close relationship between the travel behaviour of today's senior citizens and the those of future ones. The growing number of senior citizens in the world. e.g. China, Hong Kong, Japan, USA etc. countries. Global senior citizen tourism market will be based solely on demographic predictions about the future of the population's age structure. However, many of these seniors won't only live longer but will be fitter and more active until later in life. Many of the will also have plenty in life. Many of them will also have plenty of time and money to spend on travel. So, will these new seniors behave like today's senior citizens? Will they adopt the same travel behaviour as the previous generation or become a new market of oldies for the leisure and tourism indudtry? However, to determine the actual number of senior citizens who will be travelling and to sought to evaluate and specify certain difficult to predict the actual numbers of senior citizen to any country. However, they can be based on the implicit assumption that there is a close relationship between the travel behaviour of past, present and future seniors. But is this a valid assumption? As the reiseanalyse travel analysis survey, which was conducted in Germany every year, offered some interesting data possibiltieis. It was designed to monitor the holiday travel behaviour, opinions and attitudes of Germans and has been carried out since 1970 year, questions in the questionnaire. Data are based on face to face interviews, with a representative sample of more than 7,500 repondents, the interviews being carried out in January each year. All results refer to the average for the defined generated, which ranges generally over ten years. The group of people then at the age of 60 to 69 age is described. This corresponds to the same generation ten years ago, when they had an age of 50 to 59 age. When this methodological approach is not necessarily very sophisticated, it does have the important advantages of being cost effective.

Psychological method to predict travel behavioural consumption.

On the psychological view point, I think individual traveler's character will have those kind of personal characteristics. First, simplicity searchers value above everything ease not transparency in their travel planning and holiday making, and are willing to avoid having to go through extensive research. Second, cultural purists use their travel as an opportunity to immerse themselves in an unfamiliar looking to break themselves entirely from their home lives and engage. Sincerely with a different way of living. Third, social capital seekers understand that to be well travelled is a personal quality, and their choices are shaped by their desire to take maximum of social reward from their travel. They will exploit the potential of digital media to enrich and inform their experiences, and structure their adventures always keeping in mind they are being watched by online audiences. Finally, reward hunters seek a return on the investment who make in their busy , high-achieving lives. Linked in part to the growing trend of wellness, including both physical and mental self improvement who seek truly extraordinary and often indulgent or luxurious' must have experiences.

Why needs to know the personal character of individual traveler's characteristics. Because if travel agents could feel which kinds of individual traveler's character, then who can predict which kind of travel package to design to them more easily. For example, how to determine future travel behaviour from past travel experience and perceptions of risk and safety? We need to concern that the influences of past international travel experience, types of risk associated with international travel and the overall degree of safety feeling during international travel on individual's travelling experiences likelihood of travelling to various geographic regions on their next international vacation trip or avoidance of those regions, due to perceived risk. Because individual traveler's experience of safety risk degree to the countries, it will influence who chooses to go to the countries/country to travel again.

Why travellers avoid certain destinations are as relevant decision making as why who choose to go to the country(countries) to travel. Perceptions of risk and safety and travel experiences are likely to influence travel decisions; efforts to predict future travel behaviour can benefit to individual tourist's decision making. As Weber & Bottorn (1989) defined risky decision is as "choices among alternatives that can be described by prodability distributions over possible outcomes" (p.114). Some psychologists judge subjective perceptions of physical reality, i.e. image of a particular tourist destination, whereas value judgement refers to the way individual rank destinations according to whose attributes. i.e. attractiveness, safety, risk etc. factors to form on overall image. So, if the individual traveler had unhappy and worried and unsafe experiences to go to where the place(country) to travel during whose vacation time before. Then, this negative travel experience will influence who is afraid to go to the place (country) to travel again. Risk of place, country, destination or region means the danger is relatively high to the place, ie. increasing in airplane accidents, crime or terrorist activity targeting citizens of potential traveler's nationality or the probability of occurrence is great , ie. recent occurrences involving travel regions/destinations under consideration or effective actions to control consequences exist. i.e. selecting safe regions and destinations, taking extra precautions when traveling to risky destinations. These risk factors will influence the individual traveler who chooses to cancel travel plan to go to the country again.

Another interesting research, how to predict behavioural intention of choosing a travel destination, which has focus of toursm research for years, but the complex decision making process leading to the choice of a travel destination has not been well researched. The planned behaviour model using its core constructs, attitude, subjective norm and perceived behavioural control, with

the addition of the past behavioural variable on behavioural intention of choosing a travel destination.
Understanding why people travel and what factors influence their behavioural intention of choosing a travel destination is beneficial to tourism planning and marketing. Understanding travel motivation is the push and pull model. The idea of the push and pull model is the decomposition of an individual's choice of a travel destination into two forces. The first force is the push factor that pushes an indvidual away home and attempts to develop a general desire to go somewhere else, without specifying where that may be. The second force is the pull factor, that pulls on individual toward a destination, due to a region specific travel location or perceived attractiveness of a destination. The respective push and pull factors illustrate that people travel because who are pushed by their internal motives and pulled by external forces of a destination. Nevertheless, how push and pull factors guide people's attitude and how these attributes lead to behavioural intentions of choosing a travel destination have rarely been investigated. The decision making process leading to the choice of a travel destination is a very complex process. The planned behaviour model is as a research framework to predict the behavioural intention of choosing a travel destination. The model based on the three constructs of attitude, subjective norm, and perceived behavioural control (Fishbein & Ajzen, 1975).

In conclusion, the factors can influence travelers who decide to choose to travel the country, which include personal safety was perceived to the highest motivation factors among the important factors which include, scenic beauty, cultural interests, friendliness of local people, price of trip, services in hotels and restaurants, quality and variety of food and shopping facilities and services. The factors include both push and pull. Push factors include knowledge, prestige, and enhancement of human relationship etc., whereas, the most significant pull factors include high technologic image, expenditure and accessibility etc. For example, Japanese travelers visiting Hong Kong. Push factors are such as exploration dream fulfillment and pull factors are such as benefits sought, attractions and good climate city. It will be the factor of future travel patterns and motivations of sub-cultural and ethic groups for Japanese choice to go to Hong Kong travelling.

Bibliography

Backman, K., Backman, S., Uysal, M. And Sunshine, K. (1995). Event Tourism : An Examination Of Motivations And Activities. Festival Management And Event Tourism, 3(1), 15-24.

Fishbein, M., & Ajzen, Z. (1975). Belief, Attitude, Intention And Behaviour: An Introduction To Theory And Research, Boston: Addison Wesley.

Hsu, C.H.C., Cai , L.A., Li, M(2010). Expectation,
Motivation And Attitude: A Tourist Behavioral
Model. Journal Of Travel Research, 49(3),
282-296. http://dx.doi, org/10.1177/004728750
9349266.

ICT Information And Communication Technology Switzerland, 2005. ICT Fakten (ICT facts).
Available from http://www.ictswitzerland.ch/de/ict%2fakten/factsfigures.asp(retrieved Dec.12, 2005) in German.

Lind, (2001): Befolkningen, Familjen, Livscykeln- Och Ekonomisk Tillvaxt. Institutet For Tillvaxtpo-litiska studier/Vinnova/Nutek.

Lohmann, Martin (2001): The 31 st. Reiseanalyse-RA 2001. Tourism: vol. 49, no.1/2001;pp.65-67, Zagreb.

United Nations Population Division (2001). World Population Prospects: The 2000 year Revision, New York.

Weber E.U., & W, P.Bottom (1989). "Axiomatic
Measures Of Perceived Risk: Some Tests And extensions." journal of behavioral decision making, 2 (2): 113-31.

Artificial Intelligent In Road Transportation Strategy

- How artificial intelligent vehicle may interact intelligent transportation tools

Can artificial intelligence (AI) and machine learning (ML) be used in the search for new " consumption" behavioral type variables that affect consumer individual or transportation service organization individual different transportation tools choices, such as road or sea or sky transportation tools? Can artificial intelligent vehicle may interact intelligent transportation tools market development?

Consumers usually have bargaining and on risk choice when they are already shopping, such as who need to accept to use any (AI) new technological products to replace human traditional behaviors, such as intelligent non-manual driving transportation market, e.g. cars are needed to be driven by human drivers on road, but it has bargaining and on risky choice, when non-manual (AI) vehicle buyers who need to depend on non-manual artificial intelligent (ML) system assists them to drive their cars on the roads.

So, any non-manual driving auto car buyers must need to believe (AI) non-manual driving vehicles (ML) systems can make accurate driving judgement to reduce or avoid any traffic accident occurrences more than human drivers' driving judgement when the (ML) systems are driving their cars on the roads. Then the intelligent vehicle manufacturers will have possible to sell their non-manual driving vehicles success.

This is the first reason or idea influences consumer individual choice to buy any kinds of (AI) non-manual driving vehicles, when consumers believe (ML) systems are more safe and make more accurate judgement to compare human or computer systems, when they are sitting in one non-manual auto driving vehicle on the road.

The another second reason or idea is that some common limits on driving consumer prediction might be understood as the kinds of errors made by poor implementation of machine learning.

Supposing driving consumers believe (AI) machine learning ability is worse to compare to human learning ability. It will also influence driving consumers do not accept to use any (AI) non-manual auto driving vehicles to replace every driver is essential on driving by himself/herself on the road. The third idea or reason is that it is

important to influence driving customers believe how (AI) non-manual auto driving technology is used in them can both overcome and exploit human driving skill and safe limits and raise more auto driving safe judgement to compare human driving safe judgement.
However, how to predict any kinds of (AI) non-manual driving vehicles future consumption effort, due to different kinds of (AI) non-manual driving transportation vehicles which have different unique functions and designs to be used by different kinds of road transportation or driving demand of consumers. For example, lorry drivers need non-manual intelligent system can help them to drive fast, but safe to assist them to transport cargo to arrive destinations from their factories or offices. Otherwise, private car driver expects whose (AI) non-manual driving vehicle can auto drive to send to whom to arrive destination in safe way and non-too fast and non-too slow speed in order to avoid accident occurrences.
So, a different road intelligent consumer demand is to define whose individual driving behavior and driving habit and driving attitude and driving judgement and driving speed demand to decide how to design whose intelligent vehicle to satisfy those driving demand more generally, as simply being open-minded about what variables are likely to influence every consumer economic choice, when who decide either to buy any kinds of (AI) products or not to buy any kinds of (AI) products to replace the different demand of consumers their different (AI) useful demand.
Hence, for these three (AI) products group of stakeholders, such as home (AI) consumer group, firm (AI) consumer group and government (AI) consumer group . These consumer groups may consider whether different kinds of (AI) products can give what is special beneficial interest to them to use. These variables can be measurable properties of choices to influence them to choose to buy any (AI) kinds of (AI) products to use, e.g. psychophysiological, biological, social influences, consumer's wealth, moods and personality, (AI) product price etc. variable factors which will influence them to decide to attempt to buy any kinds of (AI) products to use.

If behavioral economics is as open-mindedness about what variables might predict. Then , (AI) machine learning system is a way to do behavioral economics because it can make use of a wide set of variables and select- which ones predict.
In behavioral economic view point, when general consumer overall demand to the product is much than the other similar (AI) non auto driving vehicle products, such as any kinds of (AI) non-manual auto driving vehicles and any kinds of manual driving vehicles case, then any kinds of (AI) non-manual auto driving vehicles will be more attractive to cause many manual driving vehicle buyers choose to buy (AI) non-manual auto driving vehicles. Hence, it seems if any kinds of (AI) non-manual auto driving vehicle products can make more attractive variable efforts to influence overall driving consumers to feel that they have more needs to drive non-manual auto vehicles to compare more than driving manual driving vehicle.

What is the main variable effort to intelligent vehicles to attract driving consumers to choose to accept to drive them ? However, I believe that (AI) machine learning system is a main factor to raise overall driving consumers' acceptances to drive it to replace manual driving vehicle. If it can persuade or prove (AI) machine learning system ability and judgement effort is more accurate than human or computer learning effort or judgement effort, then it is possible that any kinds of (AI) non-manual driving vehicle products will be accepted to drive on the road in popular.
Machine learning system is able to find prediction value in details of how the bargaining occurs. This discovery is the beginning of the next step for driving consumer individual driving behaviors or driving habits. It raises questions that include: What variables predict to influence driving consumers to change whose driving habits or driving attitudes? How can driving consumer individual emotion, face-to-face talking with whose friends when they are sitting in the non-manual driving vehicle to influence whom driving habit or driving attitude to be changed ? Do driving consumers consciously understand why those habit driving attitudes variables

are important when they are sitting in one intelligent vehicle? Can (AI) driving machine learning methods capture the effects of motivated cognition to influence driving consumers decide to buy any kinds of (AI) non-manual auto vehicle products more attractively. So, it seems (AI) driving machine learning method is a main variable factor to influence driving consumers to feel who have more confidence to drive them more than any other kinds of similar manual driving vehicles on the road.

Consequently, (AI) driving machine learning system will be one important psychological method to influence driving consumers to choose to buy (AI) auto driving vehicle products to replace manual driving vehicles. The reason is because human and driving machine learning system both which will have limited variable factors to influence general different countries (AI) driving consumers' need desire to be raised.

- Why can (AI) driving machine learning system main factor influence driving consumer individual desires ?

Driving consumer expectations are hard to measure or predict driving attitudes and driving behaviors in (AI) non-manual driving vehicles market. Artificial intelligence is another kind of computer science development to apply intelligent vehicle market. Why do driving consumers feel need to buy any kinds of (AI) auto driving vehicles to drive to replace manual driving vehicles on the roads? What are (AI) auto driving features different to manual driving features?

(AI) is the recreation of cognitive functions in computers; it enables machines to perform tasks like humans and perhaps even better than human. In the real world, scientists develop the technological singularity, in which a superintelligence emerges with unfold human consequences.

Professionals in many industries are intensely interested in the specifics of what (AI) can do today, and how can it helps. They are considering the impact of applied (AI), in which computers are used to address a particular problem, extracting and utilizing patterns found in large volumes of data. Of all (AI)'s subfields, machine

learning is attracting the most attention. I shall explain why (AI) machine learning system is the main factor to lead consumers feel need to buy any (AI) products to use. Such as below:
For smartphone, fraud detection to medical diagnosis etc. applied (AI) technological products examples. (AI) machine learning systems can help any one of these products to do any exceed general computer learning systems which (AI) learning systems can do any skills to supply (AI) users to use to compare computer learning systems can not do any skills to supply compute users to use. It seems that (AI) machine learning system is the unique feature to attract consumer consideration in technological product market.
An term for different types of learning, and can be accomplished using different techniques. This has led to a perception that all marketing teams should have (AI) to bring a unified personalized customer experience, when consumers choose to buy any (AI) products to feel what are the different or unique characteristics to compare general computer products. Such as (AI) product has this unique machine learning characteristics, we can predict (AI) and machine learning is connected to influence consumers to feel needs.

Furthermore, over the same time period, and in contrast to predictions for roles in many industries. (AI) won't take the place of marketers and merchandisers themselves although it is already a new value to analytical and strategic marketing skills to persuade consumers to buy any (AI) products. It means different kinds of (AI) products will have different machine learning effort and unique characteristics to attract consumers to choose to buy them to use. Such as, when intelligent vehicles need have unique road driving or sea transportation or flying machine learning system when they are applied on these three kinds of transportation tool aspects. They need have good response safety driving and immediate response learning systems to avoid any boats or air planes or vehicles to crash to them to reduce accident occurrences immediately on any one of either road or sky or sea journey environment.

What is the reason why (AI) driving machine learning system can influence good at making sense to driving consumer desire? Only humans (drivers) , preferably experienced, well informed humans can understand their driving customer needs and decide how to design or reengineer any (AI) intelligent vehicle product functions. (AI) intelligent vehicle can give these professionals the means to do this better to compare manual driving immediate response control function when any vehicles are driving or they will stop immediately to close / near to them in order to reduce crash occurrence on the road, and then maximize relevance through real-time customization of the non-manual auto vehicle driving user experience.

For example, as ever, senior decision makers need to be informed, decisive and results-oriented or risk losing out. Harvard Business Review indicated : Over the next decade, (AI) won't replace managers, but managers who use (AI) will replace those who don't. Such as intelligent vehicle won't replace drivers, but drivers who use intelligent vehicles will replace those who can not control how to drive their vehicles in the most safe way. So, (AI) driving machine learning system will have possible to do any drivers' (human's) driving judgement, driving analytical mind and driving effort to be more accurate than manual driving skills. Such as how to control to drive the intelligent vehicle in the most safe way. It is general manual driving skill can not achieve to drive in the safe way.

For another (AI) digital commerce example, (AI) and machine learning are the most exciting developments in marketing and merchandising to be applied to digital commerce, such as making better decisions through trend and cluster analysis, deploying product and content in mutually reinforcing combinations, increasing customer engagement and satisfaction in real time.

Hence, the key attraction in digital commerce circles is that machine learning is designed to be self-optimizing. Optimizing for revenue example will surface are increasingly profitably selection of products (within the brand parameters selected).

When to apply (AI) capabilities and what value (AI) is delivering for

customer and company like. Unlike any technology before it, (AI) is analytical and predictive capabilities offers the prospect for each and every individual. It can maximize real time and engagement. Effective tailored (AI) technology, such as digital experience cloud technology is available now. And once integrated, (AI) starts learning and delivering incremental value from day one. So (AI) could transform the digital experience to any business organizations.
Hence, (AI) driving machine learning system can be applied to road driving skill aspect. When intelligent vehicles are invented to own the most safe driving judgement skill and they can know when either they may auto drive fast speed, when they are feeling to know when there are not many vehicles are moving close/near to them or when they need auto drive slow speed, when they are feeling to know when there are many vehicles are moving close/ near to them. Then driving consumers will have more confidence to choose to buy any kinds of intelligent vehicles to replace manual driving vehicles to drive on the roads.

● Non-manual driving transportation tool market development

If Non-manual driving vehicle manufacturers expect their (AI) automatic vehicles can attract drivers to buy. I feel them to need to consider how (AI) driving machine learning system can achieve these requirements in order to satisfy manual driving vehicle drivers' requirement to change their traditional driving habit to choose non-manual driving needs. It means (AI) driving machine learning systems can help them to drive vehicles to replace manual driving vehicles on the road. This is the main factor to influence car buyers choose to buy intelligence driving vehicles replace to manual driving vehicles. I believe (AI) non-manual driving vehicle machine learning systems, need to be designed as below:

(1) Improving driving safety by preventing accidents from happening.
Every year, drivers are facing a large number of casualties, due

to traffic accidents. The amount of killed and injured road traffic related accidents is increasing every year. The real cost of an accident can go well beyond the limits of immediate material destruction, and is impossible to evaluate.

Hence, researchers and car manufacturers are looking for solutions in order to reduce the amount of accidents. They already developed a considerable set of technologies in order to decrease the amount of casualties. Most of them (like airbags, seat-belts, anti-lock systems, shock absorbing car bodies) are efficient in decreasing the impact of an accident, and in protecting the passengers of the cars. The technologies already saved a lot of lives, but they are rarely able to avoid accidents because they do not anticipate them. Moreover, if they are protecting in many cases, the passengers of the car, they do not prevent most traffic participants, like pedestrians on bicyclists from getting injured. it causes (AI) non-manual automatic car manufacturers need to consider how to design machine learning safety system is to prevent accident from happening instead of just reducing their impact.

This can only be possible using intelligent systems that can observe the driving environment, reason and decide if there is a danger, determine how to avoid it and act if necessary

(2) Reducing energy consumption by optimizing the driving.

Nowadays, global air pollution is serious. (AI) non-manual driving car manufacturers need to concern how to design (AI) machine learning system can reduce degree of air pollution to be the most minimum level to compare to traditional manual driving vehicles.

The reduction of energy consumption if certainly one of the main challenges. Transportation is one of the major factors in fossil energy consumption, and it is also responsible for a large amount of CO2 pollution. It is difficult to ask individuals to voluntarily limit the use of their vehicle of they do not have a strong incentive to do so. Specially in regions where vehicles are needed to drive to go to work every day. It stands to reason that if it is difficult to decrease the amount of vehicles, part of the solution is to make them more energy efficient.

Hence, non-manual driving car manufacturers need to design how to improve engines, which are more optimized and need less fuel to operate, and hybrid and electric cars have been developed and are continuously being improved. But we can go beyond these solutions that do not take into account the environment in which a vehicle is driving. A growing number of scientific contributions presented intelligent systems used in order to improve energy efficiency and reduce fuel consumption, based on the optimization of the way (AI) non-manual driving (AI) vehicles are performing. Such as recharge batteries and electric engine will be predicted the popular fuel in order to limit fuel consumption to future (AI) non-manual driving vehicles. They can reduce air pollution, consume less fuel for (AI) non-manual driving vehicles.

(3) Improving comfort by anticipating (AI) non- manual driving vehicle drivers.

Finally, another application for intelligent vehicle is the improvement of driving comfort. Car industry is very competitive market. Many potentials (AI) intelligent vehicle customers need to enjoy to sit more comfortable intelligent vehicles, who will be attracted by (AI) comfortable systems improving when driving, so part of the research in intelligent systems from cars focuses on how to improve the driving experience, i.e. make it easier and more enjoyable, more comfortable to compare to traditional manual driving vehicles.

As an example, lane keeping assistant systems are technologies that actively keep the vehicle in the lane in highways of the driven drifts out of it. Automatic speed regulation keeps the car at a certain speed without requiring to touch the gas pedal. This can be really interesting for, e.g. (AI) non-manual driving truck drivers that spend a lot of time on highways. But these technologies have a limitation in the case of automatic speed regulation, this technology can not copy of a vehicle ahead drives slower than the desired speed, or if another vehicle cuts into the lane.

This case requires the driver to have a constant focus on the road. In order to achieve more comfort, it is better of the system can adapt to changes in its dynamic environment: let the (AI) intelligent vehicle adapt to the speed of the man-manual vehicle, or autonomously change lane when requires. Again, this requires knowledge about the environment, detection capabilities, reasoning and action planning. Intelligent systems can be used in order to create more attractive and more comfortable and more safe, less energy consumption and less fuel expenditure by intelligent vehicles.

HOW DESIGNING UNDERGROUND MASS TRANSIT RAILWAY TO BRING PASSENGERS

- Designing transportation system advantages

Nowadays, transportation and economic development have close relationship. Economic development stimulates transportation demand by increasing the numbers of workers commuting to and from work, customers traveling to and from services areas, and products being moving by lorries on the roads between products and customers. According to Bailey, Mokhtarian and Little (2008) indicated "transportation route is past of distinct development pattern or road network and mostly described by regular street patterns as an important factor of human existence, development and civilization. The route network combined with increased road transportation investment result in changed levels of conveniently reflected through cost benefit analysis, savings in travel time, and other benefits. " These benefits are noticeable in increased catchment areas for services and facilities , shops, schools, offices, banks and leisure activities.

What are the crisis of neglection to care transporation system ? Why do any countries need to design road transportation system? For example, the Japan country lacks design road trsnaportation system effectively. So, the crisis of road traffic fatalities will raise and the econominc influence will be changed. The crisis indicates more than 7,000 people die annually as a result of motor vehicle crashes in Japan. Driving when under the influence of alcohol is the leading cause of motor vehicle crash fatalities in both developed and developing countries. So, alcohol is the most serious factor to raise personal risk when drivers are driving in Japan. However, a number of studies have shown that deterring drink driving is an important way to cause fatalities. There is a demonstrative need for social change in Japan.

Japan has recently strengthened its already strict laws in order to reduce the number of alcohol related road fatalities. Those deforms lowered the legal blood alochol contant limit increased, the penalties for offenders. The Japan road traffic legal needs. Any driving a motor with a alcohol limit of 0.03 or higher Japan's maximum sentence is up to 3 years imprisonment or a fine not exceeding 500,000 yen dollars. Is law impact to reduce drinking alcohol to drive in Japan? What are economic influence of the crisis of road traffic fatalities in Japan?

The rational choice theory of offending suggests that offenders are active decision makers who influence a large number of variables into decision whether or not to commit an offence. On the cost-benefit analysis, it is the punishment a possible jail, large fines worth is the reward the convenience of driving home without the expause of a taxi and innovenience to the alcohol drivers in Japan. Instead of law reforms when it detects alcohol in the air exhaled from the alcohol and other offenders and it educates children about the dangers of drinking and it also explains why alcohol driving can also threaten drivers' life when who are drinking alcohol and driving behaviour in the same time in Japan.

On the economic influence hand, implementation of the policy deregulating alcohol sales and alcohol production did not appear to increase traffic fatalities among adult or teenage males or females in Japan. We found that male adult fatalities demonstrated a statistically significant decline following enactment of the deregulation policy in 1994 year. So, Japan implement law to threaten alcohol drinking behaviour is useful. It can influence the alcohol availability and consumption, alcohol production and sales, the 24 hours operated convenience stores or liquor discount stores incomes to be reduced. Even, Japan overall GDP is also reduced from the deduction of liquor alcohol production and sale, also the occurrence of traffic accident fatalities chances will be also reduced.

The Japanese economy has entered a rapid process of liberalization since the mid-1990 year. Many sectors previously under direct government control are now regulated by the competitive market place. The Japanese alcohol beverage market has changed. The entry of cheaper import alcohol products resulted in a encouragement of alcohol consumption to Japan drinking drivers and an raising of increasing of more import alcohol products supply to Japan. Although, it is beneficial to Japan GDP growth. But it also raise the occurrence of chance to traffic accidents rate to cause alcohol drinkers to be death or hurt when who choose drinking alcohol to drive at the same time in Japan. So, alcohol import can bring more consumption, but it can also raise many traffic accidents occurrence in Japan in the same time.

In conclusion, alcohol is not good for health to drink when the consumer often buys alcohol at drink habitually. So, if many Japanese, including the alcohol driving consumers and the alcohol non drinking consumers both who often buy different countries alcohol to drink daily. It will cause their bodies to be unhealth for long term in Japan. It is possible to increase Japan's government's medical expenses to assist the low income or poor people in the future. So, although alcohol import can raise Japan GDP growth in the short term, but it also raise Japan government's medical expenditure to the low income or poor Japanese long term in the future, So it's economic benefit will not good in the future if Japan still import much alcohol to sell in its country.

Many commercial users depend on road transport facilities, with movement of products and services from place to place on the roads, aspect of global and urban economic survival. Hence, developments of various transportation modes have become important to physical and economic developments. For example, urban locations with such relative advantages are found where different transport routes with high degree of connectivity, within the intra and inter urban road networks. On similarly, commercial activities like banking, retail/wholesale businesses and professional services can take advantage of nearness to concentration of activities attracted consumers service providers. This partly caused increase in demand for commercial space and its effects on commercial property values along commercial roads can be rose. However, some countries' roads need to provide pedestrian movements more than the businesses activities, e.g. shorten the time of lorries parking on the road to let pedestrian movements on the narrow road. If the country government did not consider the roads need to let more pedestrian movements or shorten the time of lorries parking on the road. It will cause traffic jam or traffic density of the individual roads. Hence, governments need to concern the locations of commercial property buildings and the relationship between the explanatory variables of the design road networks.

What are construction of roads design networks benefits? In fact, construction of roads increased substantially with the opening up of residential environments that also is getting much benefits from increasing demand for spaces in commercial properties. Many private companies, retail stores, commercial banks aggregate in the main roads of cities, which get advantage of opportunities afforded by locations near central of cities to attract many pedestrians concerning their businesses existence. This led to high concentration of vehicular and pedestrian movements. Specially along the access main roads in the central of cities. The main roads exhibits linkages to form networks of minor routes along which commercial properties locate. If commercial users are displaced residential users, causing sites to be at the highest and best uses with increases in the values of commercial properties. However, it seems road network development is affected by the compact nature of various routes that sometimes causes volume of traffic jam. Thus, demand for transport can't be treated solely as a derived demand road. Improved main and minor roads access an city or rural areas is a necessary (but not sufficient). Precondition for increased productivity, the UK Standing Advisory committee On Trunk Road Assessment (SACTRA, 1999) noted "various ways in which transport can affect economic growth, for example benefits include through reorganization and rationalization of production, distribution and

land use: reducing labor costs by expanding catchment areas etc."
What is land use and road transport design system relationship? Land use refers to the whole range of human activity and of the built environment, and to some aspects of the natural environment. This is a way relationship between land use and road transport. Governments need to design how to use land and how to design road transportation systems. e.g. where are built the main roads and/or where are built the minor roads are the most suitable locations in the cities or rural areas ? If the main roads is located in the not suitable locations at the centers of the cities or rural, it will case the increasing traffic volumes and levels of congestion, including air pollution, noise, ground water pollution from run-off , loss of soil functions and loss of bio-diversity to natural environment. By influencing the spatial structure of locations in the urban environment, so land use planning can help to mitigate any negative effects resulting from land use changes.
Modelling and land use transportation interactions has become an important aspect of road design transport planning. On the one side, for example, design roads in urban centers, it can increase land use and it can also reduce employees or students catching buses or driving cars' time spending to go to workplaces or schools users. Hence, the land use and roads designing transportation can give benefits to residents and employment people to reduce time to wait buses or taxies etc. public transportations to go to workplaces or schools or shopping centers etc. anywhere. It seems to assist bus companies or taxi drivers to earn more income, On the other side, designing urban transport systems is also important . Increased densities mean more destinations become within convenient walking and cycling distances and consequently the use of these modes tends to be higher. Also in dese cities public transport systems are able to offer higher levels of service and operate more economically, when the provision of sufficient road space to meet potential demand becomes impractical. It aims to reduce the danger of driving or walking in urban areas. The transport modes (that is walking, cycling, public transport) and the extent of car dependence

is less, due to driving users dependency is less on rural roads. Hence, building main roads can concentrate on designing convenience to pedestrian walking to close to their houses on the streets. However, poor transport design and land use can cause to spend too expenditure not only transport costs on governments and transport users both and also the costs of providing other services. These include the usual utilities and also education and health services as well as negative externalities , such as greenhouse gas emissions. Most such studies concluded that there are significant financial and economics cost advantage of inner city redevelopment compared with fringe development.
However, such policies won't necessarily be successfully, in particular because of the two ways road problem, they may result in additional private investments and employment opportunities flowing into the region, buy may equally result in population and employment opportunities flowing out of the target region because of the improved access to other centers. Hence governments need to analyze how to arrange the land use to assist the property developers to choose where are the suitable locations to build offices or factories or shopping centers or houses at capital or urban cities to adapt to whose the growth of living population. For example, to judge where the land use whether where main roads or junior roads are built where are the suitable locations to satisfy the lorry drivers to park their lorries are the safe locations ; to design the minor roads to let the pedestrians to feel no danger to walk on the streets when the cars are driven to near to the streets on the minor roads. Thus, the factor of choosing where the land use to design the main or minor roads areas, sizes and lengths and of the minor or major roads can influence the drivers and pedestrians feel safe or dangerous when who are driving whose cars on the roads or who are walking on the streets to arrive the offices, schools, cinemas, church, houses etc. destination.

Designing road transportation networks how to assist economic growth ? I feel it is not all transport investments will be equally effective in enhancing economic growth. Designing road transport investment is a necessary, but on its own not sufficient requirement to earn significant economic growth at either a national or regional level. There are conditions under three categories: economic conditions, investment conditions and political conditions. In fact, although in some circumstances, transport investment may be a necessary condition for enhancing economic growth, it is rarely on its own a sufficient condition. Other factors including the broader policy environment, need to be present if the investment is going to be successful in addressing regional economic objectives. My some suggestions the following key aspects as being most relevant including:

a. Scale economies for example, where these dominate, lower transportation costs through improved accessibility may encourage increased concentration of firms in core regions, until the point that diseconomies set in.

b. Size of the local market.

c. Local land and labor conditions.

d. The nature and scale of transport improvements.

e. The nature of backward and forward linkages
in the country 's local economy.

In any countries, road transportation improvements don't guarantee increased economic development. To increase economic development, an improvement needs to assist any lorry drivers to drive in short trips to reduce transportation costs and shorten time driving on the road or to make transportation more reliable, e.g. reducing the numbers of traffic jams on any roads. A proper economic climate must also exist as well as other support services. With these factors to influence transportation improvements can become catalysts for economic expansion. However, road transportation improvement that intends to induce job creation, when employers need many lorry drivers to help them to transport products and to move products on the roads often. So, the employers need to employ many transportation workers and lorry drivers to help who to transport their products to send to clients, due to the transportation time is shorten and work efficiency is rasied, so the transportation times are also increasing every day when the road transportation roles are improved. On the other side, improving transportation can raise productivity when many customers need to buy many products and the lorry drivers may drive whose lorries to transport many products between factory and office or between factory to the client's home or between the shop and the client's on the road in the short time fast.

I recommend one model links in an overall road transportation network includes these four modes.

I. Maximizing use of the existing road highway system.

II. Extending or improving the multi-lane divides system local roads and connectors.

III. Continually improving the entire road highway network in response to business activities demand.

The improvement of modern road transportation successful factors include:

- How to improve the highway network

modernization includes obsolete interchanges and other segments of the road, transport network of new designs to improve the life and service of pedestrian walking streets, rebuilding certain in main or minor roads. To the extent that labor markets operate more efficiently and more jobs are created to raise economic expansion if our governments can improve road transportation system to design to satisfy business users demand when lorry drivers need to move or transport whose products on the streets, but who will not influence pedestrian are walking on the streets. Hence, excellent transportation design network can subsequent plan efforts, it can also rise economic efficiency, community and social effects, it can also encourage transportation users to attempt to drive lorries to transport products a lot of times in one day fast and who can also avoid traffic jams occurrence on the road easily. On the one side, economic development is a concept referring to the material aspects of community welfare. There are numerous factors need of development: growth in income and wealth, equitable distribution of income, decreased infant mortality rates, increased literacy rates.

On the other side, economic growth means which is sustainable increase in community income and /or wealth. (wealth is the net of resources that generate income). It seems the link between transportation facilities and economic growth has close relationship. Good transportation facilities support economic growth by lowing the transportation costs of users of the transportation network, such as roads. Direct users benefits are reductions in travel, times and fuel consumption, increased reliability and increased safety in the movement of people and products, users' transportation costs are reduced, resources are used for other purpose.

The relationship between transport and economic development occur in two directions, in the sense that (i) land use and economic development are major drivers' of demand for transport (in terms of quantity , type, location and mode); and (ii) transportation investments and other initiatives (such as regulations, pricing) can influence levels, patterns and locations of economic development. The principal role of road transportation is to provide access between spatially separated locations for the business and household sectors, for both commodity (lands transportation) and person movements. For the business sector, this involves connections businesses and their input sources between business factories and other business shops and between business and their markets. For the households sector, it provides people with access to workplaces and education facilities, shops and social recreation, community and medical facilities etc. on the roads. I feel different countries' road transportation system can be self funded in the sense that the majority of the costs of transportation system investment operation and maintenance are either paid directly by users (for example, through car operating costs) are funded initially by governments and recovered from transport users (for example, through petrol duties and road user charges). Governments' road transportation system and their use also give rise to some external costs(externalities). These include global environmental impacts (greenhouse gas emissions) and local environmental and health impacts (for example, noise partial pollution and road accident costs). The direct effects of transportation investments are to reduce road transportation time and costs through reducing travel time, decreasing the operating costs of transportation and enhancing access to destinations within the road network. A good road transportation network also needs to reduce any economic disbenefits, for example where projects reduce congestion or the risk of injury. These incremental benefits of transportation investments may be measured through commercial cost benefit analysis. Other indirect consequences of road transportation network should also be considered when evaluating effects on productivity and the spatial pattern of economic development. Good road transportation design network benefits can include lower costs and enhanced accessibility, due to better transportation links and services expand markets for individual transportation using business and improved access to input.

The economic contribution of road transportation policy can be assessed from various perspectives. These include:

- Effects on aggregate economic welfare (e.g. the sum of consumer and which is the times of cost benefit analysis, as linking to transportation productivity effect.
- Micro economic, for example, enterprise or household level productivity effects.
- Macro economics, for example, contributions to GDP investment or employment and the spatial patterns of economic activity.

One key characteristics of road transportation is split between infrastructure and operations. Infrastructure refers to the right of way on which vehicles operate, which may include ancillary facilities to ensure efficient and effective operations (for example, traffic signals, railway stations). In developed countries, are in most transportation is operated by the private cars, road trucks, the majority of bus and coach services. In long term , overall purpose, to ensure transportation system helps to develop that maximizes the economic and social benefits and minimizes harm. Hence, governments need to concern who are their main target users to use every road. Such as the road is used to near to park and leisure, or local and national economic conditions, keep clean natural environment etc. facilities to provide different benefits to different target users to enjoy to use. It seems that good transportation networks designing can influence economic activities, shopping convenience or business convenience etc. activities to cause whether the country's economic behavior to achieve close relationship successfully. Possible relationship between road networks, location attribute, demand and supply and accessibility and commercial property values of these factors which will influence different countries' concerning to choose where to build main roads and sub minor roads in different cities and rural locations. However, I shall suppose hypotheses how governments to find the most suitable places to build main roads and sub minor roads to whose cities and rural. There is no significant relationship between commercial property values and individual contributions of explanatory variables to variability in commercial property values in whose countries.

In conclusion, I suggest methods how to design suitable transportation networks to governments to build, such as it is essential to establish a technique that may be useful for determining relative accessibility of locations in the network of main roads and sub minor roads. Even, when relative advantages are determined, there is need to develop models that will be useful for predicting commercial properly values. The model may become tool for professional estate surveyors and values to change their practice of using intuition to determine relative access of locations in a road network. Similarly, there is the need to predict the supply of, demand for, and fair market values of commercial properties by developers. Hence if the cities or rural locations can attract many businesses to build commercial properties, governments can build the main roads in the locations. Otherwise, if the cities or rural locations can not attract many businesses to build commercial properties, governments can build the sub minor roads in these locations. Hence, the main roads must have high transportation valuation to let big lorries to drive and park in these main roads easily and conveniently. It seems capital cities may not influence to build the main road factors. Natural environment, commercial properties values, the lands areas size and shape and pedestrian walking numbers on the streets and lorries available numbers on the areas will be other factors to influence where to build main roads in any cities or rural in the country.

In road concept, the route network consists of primary and secondary roads, known as main roads and minor roads respectively. Main roads are usually moderate or high capacity roads that are below highway level of service, carrying large volumes of traffic between areas in urban centers and designed for traffic between neighbors. They have intersections with collector and local streets and commercial areas, such as shopping centers, petrol stations and other businesses are located along such roads. In additions, main roads link up to expressways and freeways with inter-changes in cities or rural. Road network constitutes an important element in urban development , due to urban areas have many farms, gardens, forests , so roads and building needed to provide accessibility required by different land uses and the proper functioning of such urban areas depends an efficient transport network existence. In computing des, the network indicator are used to partition road network into different parts in reasonable way. The results in number of connection to describe density differences in road networks. The parameter records how many roads connect to each road in a network. For two roads with the same length, the ones in the dense area will connect to more roads than that in a sparse area and the connection differences will indicate the density differences to some extent, so road density can also be calculated as the total length of all known roads divided by the total land area in a road divided by the total land area in a road network. Hence, governments need to consider road length to decide how to build main or minor roads to design its transportation systems for businesses activities , such as driving lorries and parking lorries and products are been moving on the streets from roads easily and conveniently. As Wikipedia Contributors (2008) indicate that "transport networks are spatial structures designed to channel flows from the points of demand to points of supply and to link the points together in a transportation system. They are useful for transport network analysis to determine the flow of people, products, services and vehicles." Hence, governments need to research whether where the shopping centers, cinemas, houses, hospitals, schools, offices, factories etc. are located, then, which need to follow these location datas to predict the cars, lorries, taxies, buses etc. of the demand numbers of transportation users to design the lengths, width and distances and the construction of main and minor roads locations and their supply numbers in different capital cities or country roads. It aims to reduce traffic jams and shorten time and air pollution as well as increasing the available spaces to let the lorry drivers to move their logistc on the road easily and reducing the accidents occurrence when the pedestrians are walking on the streets. If the vehicles can be moved on the roads easily. It will also increase time efficiency and productivity to any businessmen. Hence, how to design of the main roads and/or minor roads in any capital or country cities. It will influence any country's economic growth long time in the future.

● Underground train transportation needs to know passenger behaviour reasons

Understanding individual passenger behaviour is essential for the design MTR transportation, because who can choose to catch bus, taxi, tram, train ferry etc. different kinds of public transportation tools. Individual traveler who decides to catch which kinds of public transportation tools, it depends on whether the public transportation tool can provide real time travel information, liking link travel time schedule. So, MTR underground train needs to understand where it has terminal to give convenience to the local living areas of time travelers to choose to catch MTR easily. Although, MTR ticket fare is one factor to influence any passengers choice. But, those other factors can also influence them to choice. e.g. MTR any terminal location of convenience, short time travelling, none crowding in busy (peak) time, MTR platform waiting arrival time, none sudden MTR engineering machines broken accident events occurrence frequently etc. different factors, any one of these factors which can influence passengers who choose to catch MTR or other kinds of transportation tools.

Why route choice can influence passenger behavioural choice ? Usually, the busy time passengers will regard the route choice as a coordination problem to influence them to choose to catch which kinds of transportation tools. The route choice is as an opportunity costs to influence any busy time passengers to decide to choose to catch which kind of transportation tool which is the best right choice in the right time among of them. In the short time, for example, it seems any busy time passengers will choose to catch bus to substitute MTR underground train transportation tool, due to who feels the bus can arrive any destinations to compare other kinds of transportation tools in the most short time. However even if the MTR can either charge cheaper ticket fare to sell full day or charge discount ticket fare to sell in the busy (peak) time to compare to bus fare. It is possible that the busy time passengers will still choose to catch bus, if between the bus terminal and the another bus terminal that distance is the shorter time route to spend time to arrive destination to compare between the MTR terminal to the another MTR terminal arrival time . Also, although the busy time passengers will feel to enounter traffic jam to influence sitting or waiting bus time to be longer time in possible and who also feel MTR can avoid traffic jam problem. However, usually any busy (peak) time passengers will feel the chance of traffic jam occurrence will be less. So, the short bus route choice is more potential factor to influence the busy (peak) time passengers still to choose bus to catch.

However, if anyone wants to investigate results of day-to-day route choice which can be transferred to more realistic environment. It

is necessary to explore individual behaviour in an interactive experimental set up to ensure busy (peak) time passenger transportation behavioural choice. For example, a passenger has a choice between a main road (M) and a side road (S) for travelling from (A) to (B). (M) is faster if (M) and (S) are chose by the same number of passengers. So, this method can be researched whether MTR terminal station is located at the main road (M) or the side road (S) where is more suitable to accept to passengers generally.
Why trip time reliability and crowding factors can influence MTR passenger choice? Other problem is MTR busy (peak) time's crowding in public transportation occurrence of MTR underground train transportation tool is becoming a growth to concern as MTR demand growth at a busy (peak) time. To capture the MTR passengers benefits with reduced crowding from improved MTR public transport service and image. It is necessary a identify the relevant dimensions of crowding that are meaningful measures of what crowding means to MTR passengers. Two main influences on MTR model choice that are growing in relevance are trip time reliability and crowding. It represents a benefit-cost framework. In fact, MTR passengers can be willing to pay more expensive ticket fare, it MTR can avoid crowding and short and the accurate arrival trip time between terminals is reliable to occur. How to measure of MTR crowding, e.g. weighting the gap between the busy time, the standard (i.e. objective) and the perceived (i.e. subjective) metrics. We are not in a position to definitely map the two dimensions, which is a crucial requirement for translating objective improvements into equivalent subjective gains that then can be applied, willingness to pay estimates MTR ticket fares to obtain the additional MTR passenger benefits of MTR public transportation investment to any terminal stations. Because MTR crowding has a negative impact on passengers in terms of psychological on emotional distress. MTR passengers are willing to stand for up to 20 minutes of the service is fast and reliable. However crowding outweighed these benefits from a MTR passenger's perpective, experienced crowding leads a increased dissatisfaction. e.g. stress

and less privacy during who needs to stand up in MTR. Due to there are no enough places to supply to them to stand up in MTR. If the MTR trip time was longer time between the passenger's terminals, who will feel more dissatisfaction and it will cause who feels whether who ought need to choose to catch other transportation tools to substitute MTR next time. e.g. bus, train, tram, ferry, taxi etc. So, from an operator's perspective, the MTR service frequency or MTR size is significantly influenced by the level of ridership, which sends a signal to respond if the monitored crowding level exceeds the benchmark standard in the busy time. e.g. in the morning time or at the night time, the students or employment people who need to go to schools or offices (working places). The locations of different places between MTR terminals and crowding are regarded as a key service attribute for MTR pubic transportation along with other factors, such as travelling time and reliability, e.g. service quality, none engineering machines are broken to cause MTR stops suddenly.
Given the increasing importance of crowding on both the disutility to existing MTR public transportation users and the influence to it. MTR passenger can choose to use either the MTR public public transportation or other public transportation. It is timely to review the MTR current measures of crowding defined by transportation authorities. MTR operators ought evaluate whether they apporpriately reflect MTR each traveler experiences and perceptions of crowding in busy (peak) time. I suggest that MTR needs to buy other underground trains to supply to the busy (peak) time passengers to let them have enough seats to sit down, so who do not need to stand up in any MTR underground trains when they catch MTR underground trains in busy time. It aims to let who are willingness to pay the estimation of reasonable ticket fares to compare the other kinds of transportation tools in the busy (peak) time.

What is the crowding difference between train and MTR underground train? In fact, crowding won't be happened to brother these transportation tools easily in the busy time and non busy time both. e.g. bus, taxi, train, tram, ferry. Because passengers can not choose to stand up in these transportation tools easily, due to these transportation tools have no enough areas (spaces) to let them to stand up easily . So, the crowding will be avoided to occur in these tranportation tools usually. Otherwise, MTR will have many passengers who can choose to stand up because MTR design of length is very long and it has enough areas (places) to let passengers to choose to stand up, even there have none any seats are provided to let them to sit down. So, MTR passengers will feel more dissatisfaction and crowding easily, especial in any peak (busy) time every day.

Comparing to bus, much more diverse crowding measures are defined in the passenger rail industry. For passenger, different specifications for measuring crowding are found across countries and even within a country. For example, rail crowding measures in the UK, the passengers in excess of capacity is crowding measure that applies to all London and South east operators weekday train services at a London terminus during the morning peak from 0700 to 09: 59 , and those departing during the afternoon peak from 16:00 to 18:59 (office of rail regulation 2011 year). The overall PIXC figure is considered the planned standard class capacity of each train service as well as the actual number of standard class passengers on the service at the critical point. i.e. the location on a trains of standard class passengers that surpass the planned capacity as the difference between the number of actual passengers and the capacity of the train divided by the number of passenger is within the capacity . So, it seems train and MTR underground public transportaton tools had been encountering the crowding problems in peak time, the difference in train passengers need to wait next train or more train arrival is who doesn't plan to enter the train, when who discovers the current train has no seats to provide to them to sit down in whose trip. Otherwise, MTR passengers can choose either to stand up within the large areas (places) if who discovered there are no any seats to provide to them to sit down or who can wait the next MTR arrival in order to who can sit down. It seems MTR transportation tool crowding environment includes in waiting platform and inside of the MTR underground train. Otherwise, train transportation tool crowding environment only includes the waiting platform and the passengers will not have crowding feeling inside of the train, due to none of passengers choose to stand up inside any trains because any train inside has no enough places to let them to stand up.

How MTR can attract many passengers. On the commuter departure time choice of any reference point researching hand, the departure time decisions of communters are of fundamental importance of peak period MTR traffic congestion. However, whether on the demand side, MTR underground train congestion relief measures, such as MTR ticket fare to every terminal station needs to be charged cheaper fare or discount fare in the peak (busy) time every day. To aim to attract many passengers to choose to catch MTR Underground train public transportation tools, substitute to choose other public transportation tools in the peak time.

Over the past decades, there have been very active research efforts in the departure time problem, both in econometric modeling and dynamic user equilibrium fields. Although, these works provide valuable insights into dynamic commuter decision making, they do not identify the commuters' response to gains and losses related to whole actual arrival time to reference points who may have relative. The appliability of the reference point hypothesis of prospect theory to the commuter's departure time decision making to obtain a better understanding of how departure time choice in MTR platform during their waiting underground train arrival time. However, every MTR underground train actual arrival time and deviation variables related to reference points (gains and losses) are the key factors in the departure time choice model. How the MTR underground train of every communter's daily departure time decision can be modelled when the reference point hypothesis of prospect theory. The MTR underground train's schedule delay is defined as the difference between the preferred arrival time (PAT) and the actual arrival time (AT) for a given MTR communter. In a daily MTR commute, a commuter in the indifference band actual arrival time is an essential feature of MTR schedule study. Two reference points are the earliest acceptable arrival time and the work starting time for a given MTR platform waiting passengers. In psychological view point, prospect theory proposes that the displeasure of a loss is perceived or greater than the pleasure of a gain of the same attitude and therefore, the value function is stronger for losses than gains.

To conclude, it seems that if MTR waiting passengers need not spend long time to wait underground train arrival in platform and it can provide seats to let them to sit down in the busy (peak) crowding time. It will make them to feel pleasure, even the MTR ticket fare is not fair and reasonable to charge higher fare to compare other kinds of public transportation tools fares. So the peak waiting time factor can influence the passengers to choose other kind of transportation tools to catch easily. Moreover, MTR's two reference points are the earliest role. Similarly a loss is observed when the MTR platform waiting commuter experiences or actual arrival time which is beyond that the MTR schedule time. Due to that a MTR waiting commuter is as an early side arrival of whose actual arrival time is earlier than whose preferred arrival time.

Reference

Bailey, L., Mokhtarian, P.L. Little, A. (2008). The broader Connection Between Public Transportation, Energy Conservation And Greenhouse Gas Reduction, Report Prepared As Part Of TCRP Project J-11/Tasks Transit Cooperative Research Program, Transportation Research Board Submitted To American Public Transportation Association in http://www.apta.com/research/into/online/land_use.cfmi, accessed 17 April 2008.

The UK Standing Advisory Committee On Trunk Road Assessment (SACTRA) (1999). Transport And The Economy (Report To UK DETR). Retrieved From: http://webarchive.nationalarchives.gov.uk/20050301192906 ;

http://dft.gov.uk/stellent/groups/dft-econappr/documents/pdf/dft_econappr_pdf_022512.pdf

Wikipedia Contributors (2008). Arterial Roads In Wikipedia, The Free Encyclopeda, http://en.wikipedia.org/w/index.php?title=Arterial_road&oldid=212832640(accessed May30,2008).

- How to let passengers feel impact of undergrouund train transport to their
working time efficiency

Any countries must need road, sea and air transport to assist businessmen to transport products in local or overseas. If the country's road , sea or air transport system service quality is poor. It will influence any products transport time, speed, inefficient transport to anywhere.

How to raise the country's transport system in order to improve efficiencies to let any businessmen can deliver their products to anywhere easily,e.g. warehouses, client homes, supermarkets destination in the most short time to avoid delay occurrence to let clients feel unsatisfactory or complaint their perform their delivery services poorly. I shall discuss the factors how to improve any countrues' transport systems to achieve the most efficient way as below:

Any countries' transport systems will create economic value, e.g. demonstrate value for money, economic worth, viable commercial worth, financial affordable worth, achieveable worth. Any countries' transport systems can bring welfare value by economics. It has direct relationship to take the form of measured economic activity, i.e. GDP. The form of measured economic activity can impact on any countries' economic economic geography, locally , regionally and nationally's local GDP impacts. The welfare impacts may include: leisure time savings, e.g. the local people drive cars or catch any public transportation tools to go to any geogrpahical location's shopping centers, big gardens, swimming pools, cinemas etc. places to carry on any kinds of leisure activities.

Environmental impacts may include avoiding noise, air pollution on road transportation aspect , when the main road is only on on focus on the main city,
but the city lacks other roads to let any drivers can choose them to drive, instead of the main road in the city. Then, when many cars are driven on the busy transport
time, e.g. morning working time or night busy time between 6:00 and 9:00 AM, between 6:00 and 9:00 PM. When either many working people need to catch public transport or drive themselves cars to go to offices to work or they need to catch pubic transport tools or drive themselves cars to home. Then, the only one main road problem will need them to stay themselves cars on roads, due to traffic jam or traffic accidence occurrence problem causes when many cars are driven on the road in the busy transport time. It will influence they can not go to offices or homes easily daily, even in the busy transport time, their cars' gas need to be used much to cause air pollution and traffic noise is easily caused easily in the busy transport time on the road. When the city has only one main road for drivers in the busy transport time. So, poor road transport system can bring poor impact on economic welfare benefits arising from proved labour supply from commuting, time savings, including exchequer benefits. Consequently, the county's GDP will be fallen down, due to labour market effects which do not add to welfare value.

Whether can poor transport system impact indirectly on GDP or not on local, regional , or national economic geography impacts? Does transport lead to greater economic activity i.e. higher GDP? DO they lead to change in economic activity location? Does transport impact the existence of business location and new economic activity opportunities? The measurement on every country's transport how impacts on economic change, facilitating geographic division of labour and specialization. It can be analyzed on these general aspects:

Costs and speed of travel time (Economic value of travel time savings) . Travel time savings to users from improved transport is a key of economic value, but it has only less influence,journey time reliability is more important to business frieght as well as business travellers, network connectivity enhancements as well as business travellers, network connectivity enhancement can help people and goods travel more quickly (i.e. linked to jounrey time and journey time reliability, as well as opening new destinations and new journeys, comfort and quality service provision is relevant to public transport, e.g. detering jounreys at particular times or by certain modes (e.g. overcrowding), impact on productivity at work for commuters, safety and security , due to loss of output from workers, transport accidents occur easily. All of these issues will impact any countries' standard of living to local people (geography) , even GDP income.

Why does the direct and indirect effects of transportation have a positive impact on the economic growth and development of a country? Does it influence acccess to goods, services and employment opportunities in any regions? Underdeveloped countries must need to consider how transport system influences their economic growth. For example, the costs of transportation and production are reduced through timely delivery and enhancing the economies of scale in the production process, when the road is often traffic joam, gas cost, time waste , air pollution cost, noise has many roads, but if one lorry drivers needs drive more than one day to day to deliver goods to another city's warehouse every day. It will bring psychological pressure in terrible, when they need long time to drive on the road. They can not sleep easily because road accident will occur easily when they need to spend long time to drive lorries on the road.

So, how to solve the long driving time on road transport problem will be one issue concerns human life welfare benefit aspect, instead of economic benefit aspect. The transport system welfare worth needs to include human life worth. It is a valuable insight into the causality (ot lack of causality) between transport and economic growth and will serve to compare to any countries' national level and local geographical location level both.

In special, underdeveloped countries' public transport time whether it is long or short factor, it will influence workers their going to offices to work time. If they often need spend long time to catch buses, due to traffic jam,then it will influence their efficiences to be reduced, productive number is influenced to reduce also, because traffic jam causes they often go to offices too lately.It can influence workers' bad emotion to work every day. So, traffic jam will bring negative relationship between low efficiency and bad emotion to the workers, because they need to spend long time to wait, public transportation tools and traffic jam also influence their working emotion. Consequently, service and working performance will be influenced to poor, because long time traffic jam problem causes their bad emotion to work. It is one critical factor in the path of more widely spread economic growth and urbanization for traffic jam problem to underdeveloped countries.

However, transport system can also influence developed countries' economy. How does it influence on environmental impacts aspect from mature stage. Its business activities must raise, dramastic expansion during this period, such as underdeveloped country, US, UK. In order to acheive long term sustainable development , new demands are being placed on transport sector, such as underground mass transit rail transport , ferry, local air frieght transport, train , e.g. Japan, Fance, US high speed prior rail. Because their developed countries , business and entertainment activities needs increase, it influences high time efficient and rapid speed public transportation tools needs are also needed in societies. These new technological public transport tools invention will impact on climate, noise, human health, land use and damage to ozene layer, acidification aspects, instead of economic beneficial aspect.

For long -term sustainable development to be achieved, the various activities within developed and underdeveloped societies must be adapted to what can be tolerated by humans and by the natural environment. Transport is an activity which affects humans and the natural environment for both the development of society as a whole as well as for the mobility for the individual. For Swedish underdeveloped country example, air pollution in Swedish urban areas has beed reduced, but in many places concentrations of certain substances deiving from transport activities are still at unacceptable levels and much more has to be done. Carbon dioxide emissions and noise are examples of environmental problems demanding further efforts. Measures to limit the exploitation of valuable natural and cultural environments to protect biological diviersity are also needed. So, if Swedish still hopes to develop its tourism industry to attract many travellers to choose to travel itself country. It needs to solve environmental problems from different modes of transport are of different dimensions, such as improving its air transport to avoid cause different problems and rail transport differs in turn from road transport.

The transport problem to Swedish may include poor technological communication information to its public and purchasers of transportation and communication services as to the environmental effects of different solutions is significant in creating the demand for environmentally sound public transport service concepts. It is therefore important that such lacking high technological communication and information system is presented in as completem accurate and clear way as a method for non-monetary comparison of the environmental public transport service system aspect.

In real, it's public tranport service system is needed to be improved and upgraded in order to let travellers feel Swedish's any rail, underground train, ferry, bus , taxi etc. different public transport travelling service can provide excellent performance to serve their travelling passengers, when they need to catch any kinds of public transport tools to go to travel. They can feel convenient and comfortable to attract them to visit Swedish to travel again. Then, its tourism industry GDP income will be raised, if Swedish government can innovate any new kinds of purchase ticket equipment to install in and public transport stations to let travelling passengers feel that they do not need to spend long time to queue to buy tickets to catch ferry, train, underground mass transit rail on stations conveniently. Because long time purchase ticket queue waiting will cause travellers feel its public service performance dissatisfaction and they will complain , even they won't choose to catch the kind of public transport, even the travellers won't choose to travel Swedish again, if they feel Swedish is one developed country, but it neglects to take care about travellers' catching public transport travelling service needs.

It is one poor or bad feeing to let travellers choose to Swedish again. Hence, Swedish needs to improve its public transport service performance in order to achieve to raise their comfortable and satisfactory catching public transport tools needs to let travellers to feel. They may include efficient land use for transportation tools, comprising issues concerning natural and cultural environment, natural resources, biological diversity and aesthetics, noise reducing, public transportation energy consumption and time consumption reducing, raising public transport service facilities performance functions and other issues concerning the model. For example, Swedish government can facilitate the public transport price conparison and journey time spending comparison information gathering enquiring machines public transportation selection method of public transportation services to let every travellers can evaluate different modes of public transport when they are staying in ferry, bus, train, underground mass transit rail, taxi stations.

A travelling family can seek its sustainable transport selection system for passenger transport tool. When they touch the enquiry machine, they can compare busm ferry, train, underground train, taxi price and journey spending time from their transportation stations to another destinations. Then, travelling passengers can compare these public transport tools ticket prices, journey spending time immediately when they touch the public transport enquiring machines in stations any time. Then, they can make the most righ choice to decide whether they ought catch which kind of public transport tool to arrive the another journey destination. It is one every attractive high technological enquiry method to help any travelling passegners to choose which kind of public transport tool, it can be the most cheap transport tool at the moment in any public transport stations. So , for developed countries innovative its public transport service performance will need future passengers' journey needs daily. Hence, they can not neglect how to improve public transport service needs to satisfy passengers to feel satisfaction, if Sweden government hopes its tourism industry can raise GDP income in long time.

- How underground train MTR can let passengers to feel catching time reducing

It has close relationship between globalization and global tranport development. How globalisation impacts on the environment via changes taking place in the transport sectors. In fact, it is not clear how the relative price changes that result from openness will affect the environental composition of economic activity. For example, some countries will produce more environmentally intensive goods, others will produce fewer. On the other hand, liberalisation will raise incomes, perhaps increasing the willingness to pay for environmental improvement. These potential income effects increased outweigh the negative scale effects with increased economic activities. When combined with the positive effects with technology transfer, the net effect on local pollutants could be positive . Hence, we need to find methods to solve the problem of raising transport economic activities and serious environmental pollution creating as the same time occurrence.

Globalisation helps to facilitate greater division of labor, and to exploit its comparative advantage more completely. In longer term, globalization also stimilates technology an dlabour transfers, and allows the dynamism that accompanies economic activities to stimulate the development of new transport technologies and short time transport processes that lead to global welfare improvement.

On shipping transport industry aspect, shipping will increase ocean pollution, when international shipping activities are increasing. Trade and shipping encourages energy use in shipping is coupled with the movement of waterborne commerce. The estimates depending on the transport goods number of at-sea or in port days much increase globally every day. The energy demand of international shipping fuel sale number and domestically assigned fuel sales number also increases for global fuel usage. Estimates of ocean going ships now consume about 2% to 3% and perhaps even as much as 4% of world fossil fuels.Hence, when global shipping energy fuel usage number increases, because global shipping trading activities number increases. It will bring the environmental pollution to ocean level increases.

On air transport industry aspect, their travellers‘ catching air plans travelling needs and businesses' goods transport air delivery service needs are increasing from the requirements for high quality , fast and reliable international transport. Moreover, the networks that airline companies operate have changed often to hub-and spoke networks, many new often low -cost companies have entered the air freight market, any long time air journey is needed, e.g. Australia airline expands its one new air journey flies to UK, it needs two days flying time. It means that every flight to UK from Australia , it needs to use more fuel to fly. Then , air pollution will increase also.

On road transport industry aspect, global road transport cost and transit times, traffic jam occurrence chances also increase because when the road building number is increasing globally. So, it will cause traffic jam and long journey time spending , even fuel usage spending number is also increased. Then, accident occurrence chance is raised. Hence, global business or entertainment transport activities number increasing , it will bring much negative impact on environmental pollution, traffic jams number increases, long journey spending time increases, fuel usage number increases. Although , frequent transport activities may bring GDP income.

On transport service industy aspect, but is also brings negative influence to standard of living. It means that when transport fuel demand increases, transport activities number increases, GDP income on relative any transport activities needs industy , e.g. logistic demand needs, when lorry drivers need to drive lorries to deliver goods from one warehouse to another warehouse or supermarket or office etc. different business places on the road driving activities increase. But, it also bring air pollution , traffic noise and traffic jam etc. transport problems to road and natural environment and raises worse standard of living , bad emotion to working people or learning emotion to students , due to frequent traffic jam causes , low efficiency and productivity to workers, even student individual learning time can be reduced if they need to spend long time to wait bus, ferry, rail, underground train to go to schools , due to frequent long time traffic jam occurs on the roads to influence they can not go to schools on time often when they are catching buses to go to schools absolutely in busy transport time.

Thus, although any countries need to consider how to design their transport system, e.g. how to e.g. how to choose the right locations to build roads to let many cars can be driven available easily when the morning and evening (office and school transport busy time, e.g. 6:00 to 9:00 AM morning, 6:00 to 9:00 PM in the evening transport time usually because these two transport periods are usually , there are many students and working people need to catch any public transportation or drive cars tools to go back homes. So, enough roads number and long and not narrow road area must be needed to design in order to let enough cars be driven on the roads in the transport busy times to the countries have many big cities or have high population , such as UK, US, China, India, Hong Kong. They have many people , but drivers and cars numbers both are increasing. So, efficient road design and road number are also needed to increase in order to let drivers can transport goods to deliver, students and working people can catch any public transport tools to arrive any destinations on reads in the short time rapidly in order to avoid to spend long time transportation time and late to arrive any destinations in possible occurrence. So, any sudden traffic jam is not hoped to be caused by easy traffic accidents occurrence any time.

Hence, global efficient road transport system is needed, when global transport activities are increased, because any road logistic transport activities are increasing, they will also influence the students and working people when they also need to catch any public transport tools or drive themselves cars to go to working places or schools on the roads at the same busy transport time between 6:00 to 9:00 AM morning busy transport time and between 6:00 to 9:00 PM evening busy transport time. Because these both times will be have many students, working people , they need either go to offices or schools or go to homes. Hence, if the country had many lorry drivers need to drive their lorries to deliver goods on the roads in the transport busy morning or evening time in the same driving time on the roads. It will increase the risk to cause frequent traffic jam or traffic accident occurrence easily in possible in the country. So, any countries' governments can not neglect how to design roads and choose anywhere are the roads suitable locations to be built as well as anywhere land useful number to build road location choices in order to solve geographical traffic jams occurrence chance.

Hence, globalization of transport activities may bring geographical GDP growth, but it also bring traffic jams and traffic accidents occurrences, hearing impairment due to traffic noise, air pollution, traffic crashed, bad working emotions to workers and bad learning emotions to students, due to spending long transport time when traffic jam or traffic accidence occurs more easily.

However, transportation is an important tool if a country's progress. Rapid economic growth and increasing level of urbanization enhances a person's living standard have, it leads to a greater travel demands. Hence, governments ought not neglect have to design its roads , measure every road's length or width whether it has how many cars need to drive in morning or evening transport busy time for students, working people and delivery goods drivers of public transportation tools or private transportation tools easy driving needs in order to avoid frequent traffic jams or traffic accidents occurrences in possible.

Moreover, any governments also need to solve these issues, if they hope to develop their transport system successfully. These issues include : What mode of transportation to cost-effective in meeting a region's transportation needs to the country? How should a state department of transportation prioritize its highway delivers to maximize economic growth? What is the trade-off between additional growth in urban area and the cost of expanding transportation systems to accommodate greater growth? What effect does the expansion of transportation systems have on the need to invest in other types of transport modes? For example , the transport expansion may include the construction of additional highway segments, rail lines, runways, or additional sea, air, rail or bus terminal capacity using traditional technology; highway may include the additional of lanes to an interstate highway system; the conversion of an existing two-lane road to a four lane limited access highway, replacement or widening of bridges, and the extension of an existing road. Airport examples, include runway lengthening, apron expansion, and additional terminal gates.

On the other hand, enhancement to new transport technologies may bring efficiency of the existing highway system, examples may include intelligent highway systems, congestion pricing, intermodal freight facilities, geographic positioning systems, and instrument landing systems to mention of a few major transport innovations. So, transport policy makers need to understand the effects of these new transport mode innovations on economic development or GDP growth on transport activities growth transportation services and a more efficient use of limited land supplying scarce resources , air quality ,and noise pollution, traffic jams, long spending transport time to students, working people, entertaining people, even deliver goods lorry drivers their every day essential driving activities or catching public transportation tools needs problems. For example, the concept of intelligent highway systems needs increase trend. In simply , vehicles are being linked to each other and to traffic control devices to improve the efficiency of the total highway system. Similar types of innovations in intelligent traffic management are increasing needs for air, sea, and rail systems. The question is that whether intelligent highway systems can attribute of highways on economic development, raising on productivity of reducing highway congestion or improving pavement condition.

In fact, many developed countries' transportation system is mature. The nation has gone beyond the frontier of building, the interstate highway system and connecting most cities (markets). Tweaking the system with additional lanes and the new intelligent highway systems are useful in China, US, UK, because they have many cities. SO, road efficient traffic congestion control is needed when many students, working people, delivery goods transport people need to drive cars or catch cars on every city's roads in the transport busy time between 6:00 to 9:00 AM morning transport busy time as well as between 6:00 to 9:00 PM evening transport busy time.

However, transportation investment must be needed, if the country hoped to have good economic productivity, efficient transport service can bring good effects on the flows goods and people on roads every day when they use the country's transport system. So, any countries need to collect data, they can not be lack of enough transport information in any time that links anywhere locations of any drivers to the locations of the transport system that provide them with services in any time, e.g. every day morning and evening transport busy time, radio can report the real transport time of any roads traffic jam or traffic accident message to let drivers to listen to know whether anywhere roads are occurring traffic accidents or traffic jams or when the road traffic accident or traffic jam is solved to let the drivers can know whether when the roads can be opened to drive again. So, real time road transport message information is needed to report by radio, in order to let any drivers to know whether they ought choose to drive themselves cars on the road when they need to choose anywhere road to drive to the destination if they can know when the road has traffic accident or traffic jam occurs. They won't drive their cars on the road in the moment immediately.

On conclusion, globalization can being frequent transport economic activities. So, road , air, sea, transport service users' transport service needs are also increased. Every country ought not neglect how to innovate their transport service in order to satisfy their transport needs to achieve economic growth, efficient and short transport time spending, productivities increase, reducing air pollution, traffic noise , raisins standard of living on transport influence aspect to satisfy working people, students, entertaining people, delivery goods transport users' efficient road transport time behavioral spending aspect.

Artificial intelligent public transport how influences passenger psychology

How technology influence passenger psychology

Nowadays, robotic invention can be applied to factory manufacture, hotel, restaurant, shopping center, customer service, accounting, law document draft etc. general office tasks aspect, evem hospital surgen patient medical operation health service aspects. If future robotic non -manual driving vehicles can be invented to reach the safe auto driving mature skill stage. Any one driver begins to believe robotic, driving safe level is bette than he/she drives himself/herself car. I assume that if future robotic public transport tool drivers can replace human public transport tool drivers to drive bus, taxi, train, tram, ferry, underground train, tram , even air plane ets. different kinds of public transport tools. How non maual driving public transport tools influence our social change either to improve better ot worse? How non manual driving public transport tools influence passenger psychology, e.g. increasing any kinds of public transport tools passengers safe feeling to choose to catch any kinds of public transport tools to go to anywhere or feeling more dangerous when the passenger himself/herself chooses to sit the non manual driving public transport tool.

In past, traditional public transport tools are driven by human drivers, if one day non manual driving skills are invented to reach the most safe level, when the car owner or passenger is sitting on the non manual driving vehicle or public transport tool, the artificial intelligent driver can help the driver to control the car wheel to avoid to crash any other cars or pedestrians to o to any far places

on the roads easily. The public bus does not human driver to drive the bus, artificial intelligent driver won't feel tried, when it drives the bus long time, it does not need to leave the bus to go to toilet, to go to restaurant to eat, to go to rest room for rest, because it is one (AI) machine. So, the (AI) driver won't have negative emotion to feel angry when the bus passenger complaints its service is poor when he feels dissatisfactory to the (AI) driver bus service performance.
However, human bus driver may be complainted for unpolite or rude bus service attitude. It is common human bus driver will encounter any unreasonable passenger complain in general . Hence, when non manual driving technology can be invented to reach the most safe driving skill level, whether (AI) machine driver is the most suitable to replace any public transport tool drivers, such as bus, taxi, train, underground train, ferry, tram, even air plane to drive for future passengers service need.
In fact, any public transport tool drivers may cause traffic transport accidents, due to their careless driving to crash any other vehicles or pedestrians (walling people). Consequently, any passengers may have chance to be killed by public transport tool crashing accident. So, it seems that global public transport tools are dangerous to any passengers, when they are sitting on the bus, taxi, train, tram, underground train road public transport tools, or ferry sea public transport tools, because any human public transport tool drivers will feel tried to drive any one kind of public transport tool long time, for example when the bus driver has no enough nervous to drive the bus, he wants to sleep, due to he often needs to follow night time bus timetable to drive bus long time at night. When he often want to sleep and he is driving the bus, traffic accidents will be caused easily. So, any passenger individual life is dominated by the sleeping bus driver. His bus dirving behavior is not safe to any one bus passenger when the bus passegner chooses to catch this feeling sleeping bus driver's bus to catch.
Otherwise, (AI) non manual driver must not feel tried or need sleep often. It is one automative driving mature, it can drive any kinds of public transport tools all day, because (AI) machine drivers do

not need sleep, (AI) none human auto-driving driver can bring this important unique benefit to any kinds of public transport tools to compare human drivers. Instead of (AI) automatic driving tools' non need sleeping advantage, (AI) non-manual control auto-driving tools must not own sad, disappointing feeling , tried feeling, anygry emotion feeling when it needs to contact angry passengers every day. So, I mean that any traffic accident occurrence will reduce, when (AI) drivers often feel happy to drive any kinds of public transport tools. Otherwise, any human public transport drivers will be influenced to feel angry when they are complaint by angry passenger in any driving time easily. So, public transport traffic accident will be caused to occur easily. Althoug, it is not guarantee that it is obsolute none any public transport accident occurrence, due to crash to other vehicles, during the non manual driving (AI) driver drives the bus, tram, train, taxi, on the road, nut when (AI) non manual driving skill can be improved to the most safe driving level. I believe that non manual driving public transport tools ought be bring more safe to compare human public transport tools drivers to any one passenger individual life safety.
How non manual driving automated vehicle influences future mode of public transport service change? A survey distributed in the Netherlands in which respondents had to choose between conventional cars, public transportation for different travel distances and trip purposes. having collected information from 663 respondents, conducted a study on classic trip attributes (such as travel time, car owner self driving time and non manual driving public transport tool driving time as well as travel costs, car owner car fuel purchase expenditure and general non manual driving public transport tool fare comparison), attitudinal factors and socio-economic variables to understand future non manual auto driving public transport tools choices. The repor indicates that automated driving transport service which they defined as an automatically controlles door-to-door transport service provided by a vehicle with similar features to a conventional car, albeit driveless. Results suggest that travellers' mode preferences vary significantly

for different travel distances and purposes. They found that conventional cars and public transportation are perceived as being the least attraction altererernatives in relation to vehicle travel time and short -and -long distance commuting trips respectively, preference for passegner choice is between the non-manual driving auto car and non manual driving auto public transport tool.
They indicated that future passengers will consider how non-manual driving public transport tools whether they can bring trips are safer, faster and more efficient to let them to feel as well as traveling time and time cost is also another factor to influence them to choose to catch non-manual driving public transport tool, when they feel safe to arrive the destination rapidly. Then, future many passengers will be persuaded to choose to catch non-manual auto driving public transport tools in preference. So, if future all public transport service providers can let passengers to feel fares are reasonable price, when their non-manual driving public service transport tools, bus, taxi, tram, train, underground train etc. they can let them to feel safe to arrive destinations rapidly, they won't need worry about passengers number will reduce when human drivers are replaced by AI robotic drivers.
In fact, when one needs to drive to arrive destination in long driving time. The car owner will feel tried, bored and he/she can not spend driving time to do his/her interesting activites in his/her car, e.g. reading, listening music, watching TV, playing electronic games from smartphone, phone talking etc. personal behaviors. So, it means that future long time trip passengers may be persuaded to catch non-manual auto driving public transport tools if they believe that this kind of new non manual auto driving pubic transport tools can provide more safe, efficient, rapid, comfortable feeling to them, when they are sitting on them.
All of these may be the main factors to influence them to choose to catch non -manual auto driving public transport tools. In general, these other factors may influence future passengers to choose to catch non manual auto driving public transport tools, they may include: whether automated vehicle would drive on populated

streets better than conventional cars, whether an automated car would be comfortable entrusting the safety of a close family member, whether automated vehicle might produce fewer pollutant emissions. Because , when future non-manual auto driving vehicles are popular, many car owners will choose to buy automated vehicles to drive. So, future non manual auto driving public transport tool service providers , their competitors may be automated vehicles. If automated vehicles can let car owners to feel car prices are reasonable, they can provide safe, rapid speed, comfortable feeling to any one car owner, then he/she can sell his/ her traditional car to change new automated car easily, when global many car owners begin to accept automated cars.

On conclusion, future passengers may be persuaded to choose to buy fares to catch any kinds of non manual auto driving public transportation tools. It depends on these factors, such as reasonable fares, safe feeling, efficient and rapid arriving to destinations short time journey, comfortable and clean seats facility, free personal behavior, e.g. quite reading , listening music, watching TV , free internet provision transport environment, when future any one passenger is sitting in the auto driving public transport vehicle. So, (AI) technology will have possible to influence our future social public transportation development may bring more significant new travelling experiences and it can let global passengers to feel indeed. Moreover, it will be future global public transportation service providers, they need to consider that they ought need to change their public transport tools services in order to satisfy future passengers transport needs more easily. I conclude that future global public transport service will be influenced to change non manual auto driving public transport services by future global passegner public transport service needs within 10 years. So, nowadays, any kinds of public transport service providers ought need to spend time to research how to design themselves traditional public transport service moods to change to non manual auto driving moods in order to satisfy future global passenger individual new public transport services needs successfully.

CAN NON-MANUAL DRIVING AIR PLANES EXCITE FUTURE AIRLINE AND AIRPORT TRANSPORT INDUSTRY DEVELOPMENT

Will airline industry's ticket price elasticity be influenced by demand and supply factor

- Boeing 747 manufacturing fuel cost strategy
- How airlines and airports implement successful netwpork strategies
- Performance measurement system strategy

What factors influence cost-related management quality ?

Can non-manual driving air planes assist future tourism industry development?

- Boeing 747 manufacturing fuel cost strategy

For Boeing 747 air plane manufacture example, how it can help airline to avoid travellers number reduces. Boeing 747 air plane manufacture company how achieve air plane manufacturing strategy to reponse airline traveller number market changing. What is fuel conservation strategy to Boeing 747 air plane manufacturing firm? The cost index, (CI) feature of the flight manufacturing computer (FMC) can help airlines significantly reduce operating cost. However, many operators do not take full advantages of this powerful tool. What does CI ratio mean? The CI is the ratio of the time-related cost of an air plane opertation and the cost of fuel. The value of the CI reflects the relative effects of the fuel cost on overall trip cost as compares to time-related direct operating costs.

The equation form, CI= time cost-$/hour / fuel cosst -cents/lb

The numerator of the Ci is often called time refrated direct operating cost (minue the cost of fuel). Items, such as flight crew wages can have an hourly cost associated with them, or they may be a fixed cost and have n variation with flying time engines, anxiliary power units, and air planes can be leased by the hour or owned, and maintenance costs can be accounted for an air planes by the hour, by the calendar or by cycles . As a result, each of these items may have a direct hourly cost or a fixed cost over a calendar period with limited or no correlation to flying time.

What does this air plane fuel, cost strategy advantages to airline companies? In the case of high direct time costs, the airline may direct to time costs, the airline may choose to use a larger CI to minimize time and thus cost . In this case, where most costs are fixed, the CI is potentially very low because the airline is primarily trying to minimize fuel cost. Pilots can easily understand minimizing fuel consumption, but it is more difficult to understand minimizing cost when something other than fuel dominates. So, the cost of the CI ratio. Although, this seems straight toward, issues such among the operating locations, fuel tankering, and fuel hedging can make this calculation complicated. So , this fuel consumption cost, air plan manufacturing strategy can help any airlines to reduce air fuel useful cost and waste fuel.

However, CI can be an extremely useful way to manage operating costs. Because CI is a function of both fuel and non-fuel costs. It is important to use it appropriately to gain the greatest benefit. Appropriate use varies with each airline, and perhaps for each flight.

How low fuel situations can bring less fuel consumed benefit in the more environmentally friendly flight? Fuel conservation strategy can help airlines significantly reduce operating costs. However, many operators do not take full advantage of this powerful tool. Cruise flight is the phase of flight that falls the largest percentages of trip time and trip fuel are consumed typically in this phase of flight, which also impact trip time and fuel significantly can often be avoided through appropriate cruise planning. This fuel conservation strategy includes these characteristics. These objectives which depend on the perspective of the pilot , dispatcher, performance engineer, or operations planner can be groups into five categories , such as:

1. Maximize the distance traveled for a given amount of fuel (i.e. maximum range).
2. Minimize the fuel used for given distance covered (i.e. minimum trip fuel).
3. Minimize total trip time (i.e. minimum time).

4. minimize total operating cost for the trip (i.e. minimum cost, or economy speed).
5. Maintain the flight schedule . The first two objectives are essentially the same because in both cases the airplane will be flown to achieve optimum feel mileage.
In addition to one of the overall strategic objectives for cruise flight, pilots are often forced to deal with shortage term constraints that may require them to temporarily abandon their cruise strategy one or more times during a flight. These situations may include:
Flying a fixed speed that is compatible with other traffic on a specified route segment. Flying aseed calculated to achieve a required time of arrival at a fix. Flying a speed calculated to achieve minimum fuel flow when holding (i.e. maximum endurance). And when directed to maintain a specified speed by a air traffic control. Hence, when the air plane faced with a low fuel situation at destination, many pilits will opt to fly LRC speed, thinking that it will give them , the most miles from their remaining fuel. Also, if fuel prices increase relative to other costs, a corresponding reduction in CI will maintain the most economics operation of the air plane. If however an airline experience rising hourly costs, an increase in CI will retain the most economical operation . For this reason, flight crews typically reserve a recommend CI value from their flight operations department, and it is generally not advisable to deviate from this value unless specific short term constraints demand it.
How it can help air planes to execute for maximum fuel savings efficiently? For example, but times have clearly changed. Jet fuel prices have increased over times from 1990 to 2008 year. At this time, fuel is about 40% oa a tycpical airline's total operating cost. As a results airlines are reviewing all phases of flight to determine how fuel burn savings can be gained in each phase and in totel.
Boeing 777-200 extended range and 747-400 and (long-range, e.g. short range e.g. 717, medium range, e.g. 737-800 with winglate commercail air planes can impact fuel usage. However, flap setting must be appropriate for the situation to ensure air plane safely.

Higher flap setting configurations use more fuel than low flags configurations.
The difference is small, but at today's prices the savings can be substantial especially for air planes that fly a light number of cycles each day. Hence, top fuel conservsation strategies for flight crews include: Take only the fuel you need, minimize the use of the nuxiliary power unit, taxi use efficiencly as possible, take off and climb efficiently , fly the air plane with minimum drag, choose routing carefully, strive to mantain optimum attitude, fly the proper cruise speed, descend at the appropriate point, configure in a timely manner.
Fuel conservation is a significant concern of every airline . An airline can choose an approach procedure and flap setting policy that was the least amount of fuel, but it should also consider the trade off involves with using this type of procedure.
Hence, Boeing flight crew can earn benefit, when they fly or air planes, such as to conserve fuel and reduce noise and emissions or to accommodate speed requests by air traffic control. All of these are fuel consumption reducing strategy to airlines.

- How airlines and airports implement successful netwpork strategies

What is airline network strategy ? Network management includes route planning, scheduling optimization, airline business planning and data analysis . How implement airline network strategy, airlines need to evaluate strengths and weaknesses of the current network strategy, identification of additional potential, recommendations for adjustments, network integration, due to merger or cooperation. Fleet and capacity evaluation , analysis of estimated passenger volume over time combined with option aircraft size and frequencies for current and potential future rotes / markets; competition response, modeling of results as an impact of competitor's action and reaction; establishing best practice network managment for new carriers include: route selection, network planning, scheduling, airline business planning includes forecasting of revenues and costs. Finally, any airlines need to establish network planning, such as network optimization and development, traffic and revenue forecasting, market -and -competitor analysis, scenarios for profit-optimized networks include: hub-strategies, evaluation of aliances, cooperation includes: route joint ventures in industry environment.

- Performance measurement system strategy

Any airlines may apply performance measurement methods to design the indicates of performance. Different models and frameworks are excellence models. Any airlines hope to achieve useful performance meaurement strategy. They need to answer these kep questions: Who are the key stakeholders and what they want and need? What strategy airlines have to put into place to satisfy the key stakeholders' want and need? What critical process do airlines need if they want to implement this strategy? What capability do they need to operate and improve this process? What does cost leadership strategy mean? Competitive price is decided for customer . Indoneasia, Malaysia , India and China countries are implementing, it can provide cheaper workforces and the cost of production will be covered.

In aviation service industry, cost strategy , it much relevant to be applied cost carrier, such as Vigin Blus, Ryanair Airways are implementing . However, some airlines combined the low cost carrier and full service ,which is known as low fare limited services. Innovative marketing strategy is not only how to carry many more passengers, but also how to enable significant reduction in the costs of distribition and marketing. Such a strategy is used as a competitive strategy of low cost tariffs . For example, low cost focus strategy serve tourists or groups with certain destinations, e.g. the flights are carried out by certain airlines.

There are also tourist groups who travel to certain tourist destinations. Passengers only are transported be the tourist destinations, thus departure schedule won't be a basic need, lower price of ticket and flexible departure schedule as well as comfortable cabin are still the standard.

Another improvement of performance method, it is market

orientation, it is believed to give psychological and social benefits to the employees , in the forms of greater pride and sense of belonging, as well as greater commitment to the organization.
Another strategy improves to service performance, it is distribution management strategy . Everyone can be brought benefits. It may achieve these benefits: Raising passenger numbers, as some airlines and airports are running at near full capacity. The effects of disruption are only becoming compounded. Social media can help airlines and airports to tackle dissuption management and avoid damaging their reputation with consumers.
So, when discuption is solved. It can help airports reduce discuption cost and damaging their reputation with consumers . Innovation may include: airlines attempt to develop standard procedure for common disruption situations, responding to regulations, such as the delay rule and compensation for cancellations with faster, more proactive decisions, collaborating with air traffic control facilities, airports, themselves views of resources, identify available options.

- What factors influence cost-related management quality ?

Thus, the reduction of costs lies at the core of the low cost airline model, which aims to offer lower fares, elimating some comfort and services that were traditionally quaranteed, e.g. employed to refer to low-cost flight. The use of an airline booking system, the suppression of free-in-flight catering, the use of secondary airports connected through a point-to-point network, and the use of homogeneous fleets are only a part of the innovative choices made by low-cost airlines. However, these are main important factors to influence route structure, type and characteristics of the aircraft cost of labor and management quality . For example, if the airline's pilot, airline passenger service people, cleaners , all salaries can be reduced to employ. Then, the airport or airline's labor cost will be reduced and it can influence it's air ticket price to be also decreased. When the airline's air tickets price reduces, it will bring competitive low air ticket sale price to win other airline competitors more easily. How air ticket sale attractive prices ? The airline's air plane type and characteristics,, whether they ae comfortable to let passengers catch in their whole trip time. Whether their air planes are new or old model ? Whether their air planes' facilities can satisfy passengers entertainment need, e.g. chair can be bed to let passengers to sleep, television or movie is attractive to watch, music is soft sound etc. psychological entertainment facilities can let the airline passengers feel satisfactory or not.

Final view is management quality, such as whether the airline or airport CEO's management ability performance is either excellent or good or general or poor . They can influence whole airline or airport front -line service staffs' service performance to let passengers or airport visitors feel their services can satisfy their needs when they choose to catch the airline air planes to fly to the country to travel or work, but the country's airport staffs' service performances and airport facilities will influence their revisiting to the country's airport or rebuying the airline's air tickets again.

The final strategy is flying route strategy. I believe that whether the flying route is attractive or not, it can influence passengers to choose the airline preference. Which factors determine choice of flights on the Dhaka-London route? How could the airline be able to cope with the competitive advantages of its major rivals in this flying route? How is the competitive environment for airlines operating in this route? How could the airline sustain its competitive advantage and what can it do to gain more market share in this route?

A successful and attractive flying route design needs to amend and adjust its flying route design strategies and capabilities as the airline firm goes through its flying route design life cycle, when changes in passenger's flying route choice preferences. Hence, any airline needs to know whether what its SWOT (strengths, weaknesses, opportunities, threats) before it decides to implement which flying route(s). Because manay flying routes choicew will be influenced by its current SWOT environment situation. For organization of new flying route to the UK airport from UK , e.g. London ciry airport, the HK airline needs to know whether what its strengths , e.g. customer loyalty, its strengths can provide the most rapid . The most cheapest, the more comfortable flying feeling from HK airport flys to UK London airport or from UK , London city airport flys to HK airport, its UK , London city flying route can provide competitive fare al promotion, extra baggage allowance, airline brand image , whether is famous to UK travellers, providing online seats reservation for any UK , London flights services, airline organization size and airline brand , whether it can get travelling passegners, loyalty or confidence either to fly to UK , London city from UK city airport, or flying to HK city from UK, London city airport, providing regular UK , London flying route schedule, network presence how much UK flying route cost cutting, Uk air planes' aircrafts facilities whether are enough to satisfy HK or UK to catching the airline's air planes flying entertainment needs. Does it one new route development opportunity of direct flughts from HK airport to UK , London city airport. HK air port has enough terminals number to let facilities to provide passengers stay in HK airport when UK , London city travellers arrive HK airport? Does HK airport lacks technologica facilities to satisfy new UK , London city airport flying route development , e.g. providing enough spaces to let air fleets stay in HK airport, aircraft manintenance whether is enough? Does HK airport encounter shortages of experienced front line counter service staffs to serve UK , London city travellers in airport? When the seasonal time is holiday travelling time for UK visitors, is UK new flying route rising fuel cost? Does new entrants to develop another UK , London city flying route from HK? Has new UK , London city from HK flying route , these weaknesses , such as lacking enough resource to develop another another new flying route or no direct flight or long hour flight after the HK airline developed the new UK , London city flying route to the HK airline from HK airport. Has new Uk, London city flying route development enountered these threats: strong competiton, high interest and UK foreign currency exchange rates, raising fuel cost, economic recession decline in the UK airline industry, environmental pollution etc. issues influence UK travellers choose to travel to HK desires. So, SWOT analysis must be needed to consider in order to develop any new flying route in order to avoid servious high cost expenditure loss to any airlines.

So, above these tangible , e.g. fuel cost control , route choice, network cooperation choice and intangible, e.g. service performance factors can influence whether the airline's network cooperation strategy, route choice strategy or low cost strategy can succeed or fail. So, airlines or airports can not neglect these factors to be revised in order to achieve any strategies in success more easily.

V

AI sixth stage development

How AI technology assists human to avoid energy and food waste

House quality influences householder electricity energy consumption behavior

Can apply artificial intelligence measure the house quality how influences the householder electricity energy consumption or useful activities to be more or less? In general, house owners have both intentions for whose property. One intention is living the house by householder himself or herself or householder with families themselves. Another intention is that renting to others to receive rent income (landlord). So, in the housing market, the housing consumer includes either the property owner intents to rent to others to live for rent income aim or the property buyers intents to buy the house to be house owner to live. Does these both different property purchase intentions, which will influence the householder's attitude to use electricity energy consumption desire to be more or less, due to the householder's demand to whose house quality factor influence? This is one interesting question concerns the householder electricity energy consumption desire change to the householder, due to house's investment or house's living

intention influences to house quality factor.
How does house quality factor influence to householder electricity energy consumption desire to be more or less? Has it relationship between house quality and house investment or living intention to cause house quality demand to influence the householder electricity energy consumption desire change or demand to be more or less? Has it relationship between regional housing market living or rent investment intentions, housing quality and electricity energy consumption more or less desire? I suppose that the determinants of the residential electricity energy demand form space-heating and cooking, due to the property quality demand influence and the householder's living or rent investment intention influence both, which will influence the householder's electricity energy consumption or useful behavior when he/she/they is/are living in the house.
I argue that rent properties are not only consumer goods, but it also constitute financial market assets. It is therefore reasonable to assume that rational (rent income investment intention) investors choose to raise housing quality (e.g. thermal insulation technological installing at home, heating or cooling technology or artificial intelligent window, lighting, door opening or closing) in order to attract many people choose to rent whose house to live. The householder's aim is to achieve an acceptable return on investment (ROI) or raising rent income aim when he/she rents whose house to anyone, it is easy to attract many people to choose to pay higher rent his/her house to live in the property rent market. Moreover, the another important factor is that rents and future house sale prices of properties differ regionally (or even locally), and largely depend on housing market fundamentals, such as either the house living buyer's income levels or the house rent buyer's income levels, vacancy rates, and/or householder investor's expectations.
Thus, if the householder expects to rent whose house and raises rent to attract many people choose to rent whose house to live, who will attempt to install many new technology in order to satisfy their high quality of life need when they can pay higher rent to rent

to choose to rent whose houses to live. Their aim only achieves to raise housing quality, but any new technology will lead to increase electricity energy consumption or use in the house.

Hence, any high quality of houses will influence the householder to use or consume more electricity energy at home. It means that the householder will choose to consume or use more electricity energy at home, if he/she or the family householder demands to live more comfortable house and he/she/they can have high quality of living life at home. This comfortable living demand to the householder (property renter or property buyer) view point can explain why the better quality of house factor will influence the electricity energy consumption desire to the householder also to be more daily.

I shall indicate one home electricity energy consumption experiment, it indicated that utilizing aggregate data on regional space-heating energy consumption form over 300,000 apartment buildings in 97 German planning regions. The study applies structural equation modelling to estimate the influence of housing market fundamentals on the level of housing quality, and subsequently on regional electricity energy consumption. Consequently, it suggests that housing market fundamental explain regional differences in the housing quality.

In particular, findings show that the level of per capita income, investor' expectations about future housing market development as well as vacancy all explain regional differences in housing quality has a significant impact on electricity energy consumption.

In the way, this experiment can indicate evidence that regional housing market fundamental have a substantial influence on regional levels of housing quality and energy consumption desires to the German regional householders.

This Germany regional householder experiment found important implications for high or low housing quality of the regional property building and householders either property living or property rent intention of comfortable living feeling need factor which will influence the regional property householder electricity energy consumption desire to be raised or reduced. These factors will influence the consequence of electricity energy demand to be increased or decreased needs every day for the regional householder as well as the country's electricity energy supplier(s) can gather the regional properties whether they are high or low quality to predict the regional properties householders' electricity energy consumption supply budget more accurate. It implies that an important determinant of residential housing quality will have possible to influence electricity energy demand to be more or less for long term. IN particular, this Germany regional residential experiment can explain and find an important role in formulating assumptions about the quality factor has chance to influence the regional residential future levels of electricity energy efficiency and consumption in the country. Hence, housing developments and electricity energy firms can follow this regional residential housing quality factor to evaluate whether the regional housing market is the corresponding investment patterns as well as the energy researchers can follow the regional residential housing quality whether it is high or low housing quality factor to evaluate the more accurate models of regional electricity energy demand to any regional residential householders' houses in the country.

In consumer behavioral view point, it explains that if the country government expected many householders feel to need to spend much electricity energy or have much electricity energy useful demand or desire at home. The country government ought to encourage the country residential property or house developers choose to build many houses which have technological product installed to satisfy the regional householders' residential comfortable living need when they choose the regions to build the high quality houses to let them to live. Then, the regional householders will be influenced to consume or use much electricity energy at homes, due to they feel that they are living at high quality and comfortable and high building technological installed apartments in the country's regions. Then, the country's government and electricity energy provider(s) may be raise much electricity energy efficiency and supply and profit , due to the regional residential householders' electricity energy consumption or useful desire need is therefore influenced to be more by the regional high quality of residential houses factor. So, the regional high quality of residential house factor will have relationship to the regional electricity energy consumption and efficiency to the regional householder's houses.

Otherwise, if the country government felt electricity energy is shortage, it ought encourage the property developers build many low quality and low building technological houses to let householders to live themselves or rent to others to live in the country's different regional residential development market. Due to the low quality of properties and low technological installed to properties factor which will influence any these different regional residential householders to choose method to solve shortage of electricity energy challenge to the country.

In conclusion, to apply consumer behavioral economic theory to property development market, if property quality factor can really influence the householder's electricity energy consumption desire to be used more or less at home daily. The country's property developers can apply this factor to predict property consumer individual property buying consumption behaviors more accurate. For example, if the US property developer planned to build low quality and low technological design buildings and lesser comfortable residential houses in the region in US. Then, its residential householder target will be trended the less acceptable of electricity energy consumption property buyers to choose to buy these regional properties to live in the US region because they can only accept to spend less electricity energy to use when they are living in the houses in order to save money daily. SO, the low quality , less comfortable and low technological installed design residential houses will satisfy their living needs. Otherwise, if the US property developer planned to build high quality and high technological installed design buildings and more comfortable residential houses in the region in US. Then, its residential householder target will be trended to the more acceptable of electricity energy consumption

property buyers to choose to buy these regional properties to live in the US region because they can accept to spend more electricity energy to use when they are living in the houses in order to improve their living of quality. So, they wont's consider to spend more expenditure to use electricity energy for any technological products are installed in their properties in order to satisfy their comfortable living needs at their homes every day.
Consequently, property developers can attempt to gather marketing research concerns whether how many people who accept to use more electricity energy or use less electricity energy in order to predict they ought build how many high quality or low quality houses number in different regions more accurate in themselves countries or overseas countries property development market.

Environmental impacts of householder greenhouse gas electricity energy consumption activities

Can artificial intelligence measure whether what environment factors influence householder electricity energy consumption activities? Has environment factor relationship to influence householder electricity energy consumption behaviors? Socially, householder electricity energy consumption provides us with sources of living satisfaction , but if any sudden environment factor changes, whether it will influence householder consume or use more or less electricity energy decision at home. However, I assume householder electricity energy consumption will have a considerable proportion of the environmental impacts be influenced by our way of life and our economic decision of electricity energy consumption behavior.

What different environmental factors will influence householder electricity energy consumption decision? The external environmental factors include, for example, the country's electricity firms or government changes to electricity energy regulations, electricity energy production technologies change and business practices and government policies changing etc. different external environmental factors will influence any country's electricity energy consumption to householders' consumption desire to be more or less. It will also require changes to influence the householders to consume which kinds of electric products which are needed to be used in different electricity energy natural manufacturing resources.

Why does these external environmental factors impact householders' any behaviors to influence them to concern to use more or less electricity energy power or which kinds of electricity energy products choice at homes. How any why environmental factors impact will influence householder activities at home, such as electricity energy consumption and choice? What are the key components of external environmental factors influence householders' electricity energy consumption behaviors. I shall explain as below:

Firstly, we need to know whether what external environments are which can influence why and how householders need to change their activities to choose more or less or which kinds of energy power to be provided to them to use at home. Who is householder? Householder is an individual, family, or group of individuals living together as unit in a home. Consumption of electricity energy at home may be cooking food needs, needing have colder feeling to turn on fan or air condition at home in summer or needing have warm feeling to turn on heater at home in winter, watching television programs or listening music , playing computer games or used computers activities , reading activities and applying artificial intelligent technological tools to help householders to open or close homes' windows, doors etc. different home equipment which need to use electricity energy provisions. SO, their home activities need to turn on lighting electric tools , televisions, music machines, radios etc. different equipment which need to use electricity energy provision at home. SO, the purpose of householder consumption means consumption by individuals living in a household and it includes consumption both in and outside the home. Why does environmental impacts link to householders' electricity energy consumption? I shall focus on discussing of greenhouse gases (GHGS) energy product how any why it can influenced to householders to use.

The environmental impacts will influence this kind of greenhouse gases (GHGS) energy in the product lifecycle or delivery of the service to link the householder's energy consumption at home such as these several aspects:

Extraction and greenhouse gases production (supply number), physical distribution (delivery far long or close near short distance between the greenhouse gases manufacturing factory and the greenhouse gases supplier), resources consumed by marketing and retail activities (householder's needs to use the quantity of the greenhouse gases energy product), the greenhouse gases consumers search and purchasing activities(e.g. travel to shops, internet purchasing channel, , finding the which kinds of greenhouse gases products from internet, magazines, newspapers, radio advertisements etc. different medias,) , post-use greenhouse gases energy disposal (resale, reused or rubbish). The householder's physical behavioral impact environmental factor will influence how and why he/she chooses to consume greenhouse gases energy daily , e.g. impacts of a housing development, or a wind –farm that supplies greenhouse gases with power. So, the householder's greenhouse gases energy consumption behavior which will depend upon individual personal and subjective perspectives and value.

So, the householder's useful behavior or attitude of greenhouse gases energy product which will influence how he/she/ the family use or consume greenhouse gases energy, such as the householder individual environmental protection attitude which can impact how he/she/the family spends the quantity of greenhouse gases energy every day at home, if the householder does not expect our air or water or land is polluted , due to extraction of any natural gas resources to be manufactured any kinds of greenhouse gases products. Then, this environmental pollution issue will influence some householders choose to reduce to use more quantity of greenhouse gases products every day. Another environmental factors include the bio relates the (unsustainable) use of resources to avoid wasting much greenhouse gases energy to cause greenhouse gases energy supply shortage, avoiding the cause negative impacts of quality life , e.g. noise causing when the extraction of any natural resource from lands to the householder's house is near to the natural resource extraction land and health impacts, e.g. when the greenhouse gas householder user who often use the kind of greenhouse gas product when it is used to cook or heat any equipment to cause they to breathe dirty air at home often. These impacts can be measured in different ways include: monetary costs or loss, physical quantities of resources used or waste or pollution produced and the burden the greenhouse gases energy place on environmental resources. All of these external environment factors will impact the householder individual attitude or behavior how to use or consume greenhouse gases energy product at home.

All these environmental factors concern householder greenhouse gases energy consumer individual consumption attitude is influenced by environment pollution, greenhouse resource supply shortage challenge, greenhouse gases influence the householder's negative quality of life, negative health impacts, noise, waste money , raising economic cost to the householder which will impact whether how the householder choose to use the quantity of greenhouse gases product or the kinds of greenhouse gases products or other kinds of electricity energy products.

However, these are other external environmental factors which can impact how the householder decides to use greenhouse gas product at home. They include: the changes of energy regulation, e.g. the country government has quota number implementation to prohibit to import above the limited quantities of any kinds of greenhouse gas products to any countries. So, when the greenhouse gas energy supplying quantity is decreased, but if the country has may householders who need to buy different kinds of greenhouse gases products to be used at home. Then, the different kinds of import greenhouse gases energy products prices will be raised in possible, due to demand is more than supply in the country's greenhouse gas energy product market. Consequently, if the greenhouse gas energy price us risen above the general social acceptable level to the home greenhouse gas energy product householder consumers. Finally, it will influence them to choose to buy other kinds of gas energy products to replace the greenhouse gas energy product to use at home.

Another side, if the country's greenhouse gas energy manufacturing supplier sudden changes its greenhouse gas energy production technologies to choose to concentrate on manufacturing other kinds of energy products. Then, the greenhouse gas energy supply quantities will be only decreased, even future one day , it will cause greenhouse gas supply shortage challenge to let the country's home greenhouse gas householder consumers who can not buy enough quantity of any kinds of greenhouse gas energy products to satisfy their electricity needs at home every day. Consequently, when future on day , the country greenhouse gas energy manufacturer has none any quantity of greenhouse energy products to supply to the country's greenhouse energy householders to use at home. The, they must only choose other kinds of new energy products to replace the traditional useful greenhouse gas energy products to be used at homes.

In conclusions, these non-controlled external environmental factors can impact and influence the country's every householder consumer individual attitude or consumption behavioral change to how any why the country's householders either choose to buy much or less quantity of greenhouse gas products to use at home.

The effect of house space occupancy
and building characteristics on
householder electricity energy use

Can artificial intelligence mesaure how much space occupancy is the suitable size as well as what the most suitable building characteristics influence each householder electricity energy useful activities or behavious? In general, society believes large space size

occupancy house building characteristics factor which will influence householder use more energy at home, e.g. in summer, when the householder is living at the large space size occupancy house, who ought turn on all air conditions or fans at sleeping rooms or eating room or studying room. So, if the householder's house has two to three or more sleeping rooms. Then, he / she needs to buy more air conditions or fans in order to let all rooms' temperature to be fallen down to let he /she feel more cool comfortable feeling when the temperature is above 30 degree or more extreme hot in summer weather. Otherwise, when the temperature is low, e.g. between 0 degree to 10 degree or below 0 degree in winter weather. When the householder is living in one large space size occupancy appartment, which has thee to five sleeping rooms , even more and two studying rooms and one eating room, even more as well as every room has one heater. Then, he / she must turn on all heaters to let who to feel warm feeling when he / she is staying in the house. It brings these interesting questions.
Will large or small size space occupancy housing characteristics influence any householder often turn on heater or air condition or fan in whole house space occupancy area in order to the householder feels warmer or cooler feeling when he /she is staying in the house?
Has any space occupancy housing characteristics relationship to influence any householder to turn on heater or air condition or fan in whole house space occupancy area in order to the householder feels warmer or cooler feeling when he /she is staying in the house?
Does it bring positive relationship between turning on long time fan or air condition or heater and the house occupancy space characteristics is large or small size?
I shall attempt to give psychological evidences to explain the householder's house space occupancy area large or small size factor whether it can influence the householder choose to do long time or short time turning on heater or air condition or fan behavior in order to let he/she/the family to feel more cooler or warmer comfortable feeling when he/she/the family is staying in the house

in summer or winter weather.

Does the house occupancy space size characteristics factor is the only one or important factor to influence the householder choose to turn on long or short time fan or air condition or heater in the house to let him/her/the family to feel more cooler or warmer comfortable feeling in summer or winter weather?

I feel that it is not exact right , due to the householder's house space occupancy size whether it is large or small characteristics to influence the householder choose to turn on long time or short time fan or air condition or heater time to let him /her/ the family to feel more cooler or warmer when he / she / the family is staying at home in summer or winter weather. The reason is because that the lifestyle of living quality need is different between developed countries and developing countries. The lifestyle of living quality factor will change the country's householder's expectation about the quality of living life. For example, for Africa, Korea, China , Japan, Hong Kong etc. developing countries. On the lifestyle of living quality need to these developing countries' householders aspect, that will cause a high environmental burden when they need to often turn on air conditions to satisfy more cooler feeling when they are staying at homes in summer or they need often to turn on heaters to satisfy more warmer feeling when they are staying at home in winter. Due to if their houses are large size space occupancy characteristics and they have more than at least two sleeping rooms and studying rooms and eating rooms and toilets number. Then, these householders who are developing countries' large space occupancy size characteristics houses, they won't like often turn on heaters long time to keep more warmer in their indoor whole space area in winter or they won't like often turn on air conditions or fans long time to keep more cooler in the their indoor whole space area house environment in summer .

The reason is possible because that the developing countries' householder chooses often to turn on their heaters or air conditions or fans long time in their houses when they are staying long time in their houses and their houses space occupancy sizes are very large, it will bring the electricity energy to be used more to these developing countries‘ householders’ large space occupancy size characteristic houses. It means that the electricity fee will be also increased due to they often turn on heaters or air conditions or fans long time to keep their indoor temperature to be more cooler in summer or more warmer in winter. So, it seems that the developing countries‘ householders are living in the house whose space occupancy have very large size characteristics and more than two rooms house characteristics in the developing countries as above. Then, they won't often choose to turn on heaters or air conditions or fans long time to keep more cooler or warmer feeling in their house whole indoor space occupancy environment when they are often staying at home long time.

Their lifestyle of living comfortable feeling are lesser than the developed countries householders. Consequently, their lesser cooling or warming comfortable demand of living lifestyle factor will change their attitudes to use air conditions or fans or heaters turning on time in order to limit heaters or air conditions or fans turning on time to be shorter than the developed countries houeholders' heaters or air conditions or fans turning on time at homes. Due to the long time turnong on air conditions, fans , heaters at the developing countries‘ householders' homes, it will cause to spend much electricity energy to lead electricity fee charges to be raised to the developing countries‘ householders ' homes when they are often staying at homes in summer or winter weather. Hence, the house space occupancy large size characteristics ought not influence the developing countries householders choose to turn on air conditions , fans or heaters long time in order to let them to feel more cooler or warmer at homes in summer or winter weather.

So, the developing countries‘ house space occupancy large size characteristics householders won't be more acceptable to pay higher electricity energy fee when they are staying at homes at summer or winter weather. Due to they do not often choose to turn on heaters, air conditions or fans long time during they are staying

at homes. Otherwise, the developed countries, e.g. UK, UK , France, Germany, Swiss, Singapore, Italy etc. countries. In general, these developed countries' householders' living lifestyle quality needs are higher than the developing countries. So, when the summer or winter weather is coming, if the temperature is extreme cold, e.g. below than 0 degree or it is extreme hot, e.g. higher than 30 degee.
Then these developed countries' householders will easy accept to turn on air conditions or fans or heaters long time at home in order to keep their appartment in door temperature to be more cooler in extreme hot in door environment or more warmer in extreme cold in door environment when these developed countries' householders are often staying at homes long time at night after their day time working time or schooling time. Because these developed counties' householders' quality of living lifestyle needs or demands are higher than the developing countries' householders. So, they won't consider that they will pay more electricity fee , due to they often turn on air conditions, fans or heaters long time to let them to feel more comfortable in cooler or warmer indoor large size space occupancy environment. So, it seems that the electricity energy efficiency will be raised to the developed countries' householders who are living in the house space occupancy large size characteristics and they will be possible to pay more electricity fees during they are often staying at home in extreme hot summer or extreme cold winter weather.
In conclusion, due to the living lifestyle quality need (demand) is different between the developed countries' householders and the developing countries' householders. It will influence the householders' long time or short time spending time on air conditions or fans or heaters indoor space occupancy size characteristics environment in order to achieve more cooler or more warmer feeling in their houses. Consequently, the long or short time of turning on air conditions, fans, heaters for the developed or developing countries householders' activities factor will be more influential to compare the house space occupancy large or small size characteristics factor to influence their cooler or

warmer feeling in their houses. SO, the house indoor environment electricity energy consumption efficiency degree to the developing or developed countries' every householder house in summer or winter to the developing or developed householders in summer or winter weather , which is more influenced by the living lifestyle qualty factor to the either developed countries or developing countries householders. Hence, any developed or developing countries' electricity suppliers need to consider the building areas of property development market buyers their living style quality demands (needs) whether their living style quality demands are higher or lesser than the other building areas of property development market, they ought not consider whether the building locations of the houses' space occupation sizes whether they are large or small sizes in order to evaluate the householders will live at the building areas of property development locations ,whose electricty energy spending efficiency more accurate.

How to help low income household earners to reduce not essential electricity energy expenditure spending at homes

Can artificial intelligence measure whether which income level influences low income household accept to use the much electricity energy at home? Has it relationship between the householder income and the electricity energy needs? How to evaluate the subsidies and social tariffs to assist lower income earners to analyze household energy consumption more accurate?

Electricity energy is essential needs for every householder at home, e.g. lighting, cooking power, healthcare, sanitation, cooler or warmer temperature indoor control at home. However, for lower income household earners, it its burden when they need often to use electricity energy to supply power to any home electricity tools to do any acticities at homes. If any these countries' lower income householder earner target can not get the reasonable subsidies to assist them to solve any electricity energy tools' electricity energy poer needs. Due to their lower income leve, it is possible that to knfluence them have enough electricity supply to help them to use to cook rice and food and vegatabe to eat, boil water to drink, turning on light tools to help them to read, watch TV, listen radio, music any entertainment or essential needs at homes at night or morning afternoon time. These lower income household earners will be easy to sick , due to they have no enough electricity supply to help them to use use electric bottles to boil water or cook food to eat. Then they only drink not boiled water or not cooked food to eat at homes in possible, due to they have no enough income to pay electricity fees every month.

Hence, how to evaluate the lower income household earners' electricity fee need (demand) level in order to provide the reasonable subsidies amount to assist every country's low income household earner to help them to pay the reasonable electricity fee which is one important issue to every country's government today. It brings this question: How to evaluate or analyze or predict every lower income household individual or family earner's every month electricity energy demand (need) more accurate?

It is one essential issue to be value to consider to every country's government. Moreover, to the estent that energy subsidies must be essential to be provided by public sources to all low income household earners or that a social tariff may be designed for improving access to energy for certain low income social earner groups. Hence, how to structure the energy subsidies between energy and income levels to be better target, such public mechanisms, and to avoid regressive subsidies unfairly. For example, India and China these both countries ' income poverty and energy poverty population are the large number. So , these both countries' governments need to focuse on more aggregated effects and analyze the effects of rural electrification at the local level on the decrease in energy poverty in rural low income poverty and energy poverty householders. Therefore, every country government needs to point regressivity of the subsidy for electricity. There is room to analyze to what extent low income household earners along the income distribution demand some forms of energy, and to suggest better and fair low income targeting household earners energy subsidies supply policies.

Each government does not only consider energy issues from a social point of view, it also needs have a manner to consider a possible link between energy, hunger reduction, and food security for each country's low income household earners group. So, every government has responsibility to calculate the determinants of different sources of energy consumption at the low income houehold earner level for urban and rural both populations in order to evaluate the electricity subsidies and to test whether every low income householder earner characteristics plays a role in determining energy consumption.

In general, in the use of energy measured as that for cooking, such as LPG reduces the exposure of households to hazardous, increases the consumption of different types of foods and medicines, improves the distribution of time between household memners, enables studys with more light, reduces the use of digital computer entertainment tools at home, and moderates the use of wood as fuel, preventing deforestation. These methods are the best suggestions to help low income householder earner groups to reduce time to use electricity at homes. When they spend less time to use electricity to do any not essential activities, e.g. watching television, playing electric games from home computers, listening music. They only use electricity to turn on light read, to turn on rice cooker to cook, when they feel hungey to eat. Then, I believe that these social low income household earner groups will reduce to pay much not essential electricity energy expenditure at homes. Hence, every country government ought need to persuade low income household earners to avoid to use electricity to do any not essential activities in order to raise electricity energy consumption in long term time.

It will bring less amount of energy subsidies expenditure benefits to every country's government. Hence, the success to persuade any countries' low income household earners to reduce to spend much time to do any electric entertainment activities of consumption behaviors at homes often. This is the most efficient and the most successful energy subsidiary method to help them to reduce electricity energy expenditure when they are staying at homes. Hence, if any country government expected the low income household earners can continue really reduce electricity energy expenditure, they need to learn to do the meaning essential activities which are needed to use electricity at home habitally. Then, they can change their electricity useful entertainment living habit, e.g. using computers to play games, listening music, watching television entertainment habits at homes to cause essential daily needs of electricity useful living habit, e.f. using cookers to cook rice or cook food to eat, turning on lights to read , turning on heaters to bath, turning on air conditions to keep cool temperature or turning on heaters to keep warm temperature at homes. Consequently, they won't need to pay much electricity expenditure at home, due to their waste useful electricity entetainment living habits have changed to do any essential useful electricity activities at homes.

Another kind of method to reduce the determinants of energy demand to the low income householder earners. The governments can persuade them to consider the variation factor can influence their electricity energy expenditure are increased or decreased at homes. It is not the electricity or gas price is increased from the electricity suppliers. It is that their bad living habits of waste electricity or gas to do any not essential activities at homes. e.g. the householder often turn on light tools to read or listen music or watch television in whole night, he/she ought need to sleep at night, but he/she does not go to bed to sleep in whole night. He/she chooses to turn on light to do these activities. Then, he/she will waste much electricity at whole night. Also, some householders like to bath more than half hour, even one hour, when it is winter, they need to turn on heaters to provide electricity to cause the bath room has warm water to provide to them to bath, Their long time bathing behaviors will be also waste electricity or gas energy from long time heating in bath rooms. So, they need to change their waste electricity consumption living behaviors at homes.

So, I suggest that some low income household earners will need to

be taught to change their bad using electricity enery living habits from governments' public relation promotion in order to change the low income household earners' bad or incorrected useful electricity or gas living attitude to achieve and to avoid them often to do electricity or gas energy waste behaviors at homes. So, different countries' governments need to teach them how to do the correct or right electricty or gas useful activities (living habits) or let them know or feel how to use their electricity or gas which can help them to reduce to waste the not essential extra electricity or gas energy. Consequently, they must reduce electricity or gas expenditure as well as electricity or gas shortage challenge won't be caused by their electricity or gas useful waste behaviors (activities) at homes.
In conclusion, energy subsidies method is not the best solution to help low income household earners to reduce to use electricity or gas energy. Because it is only short term benefit to reduce their electricity or gas expenditure at homes. The best solution is that to let them to know or feel why and how they have responsibilities to change their incorrent or wrong electricity or gas consumption bad habits in order to avoid global electricity or gas energy is waste to be used, even it is caused shortage from householders' energy waste behaviors.

Factors influence householder energy
efficient consumption behaviors
at homes

Can artificial intelligence measure how householder uses energy level at home ? What factors can influence householders how to use energy in efficient way at homes. It depends on different countries householders' living habits to cause their choices to use energy efficiently at homes. In general, global householders energy every day consumption or use aims include cooking, heating, and cooling or warming rooms, lighting , water-boiled use and computer playing games entertainment etc. activities at homes every day. Some activities are often essential at homes, e.g. cooking, cooling or warming temperature in rooms, lighting , water-boiled use. So, their activities must not avoid to use energy at homes often. Otherwise, some activities are not essential at homes, e.g. playing entertainment games from computers, cooling rooms in summer, listening music, watching television etc. these activities. The householder can choose either to use energy to turn on these equipment tools or not to do these non essential activities at homes often. In general, householders rely on energy to make ourselves lives comfortable, productive and enjoyable. However, global householders need to learn how we can use energy resources wisely because global every householder has responsibility to manage resources includes: reducing total energy use and using energy more efficiently in order to avoid energy shortage crise occurrence. The choices are make about how we use energy, e.g. turning machines off when not in use of choosing to buy energy efficieny appliances will have increasing impacts on the quality of our environment and lives.

Energy conservation includes any behavior that results in the use of less energy. Energy efficiency involves the use of technology that requires less energy to perform the same function. For example, a compact fluorescent light buld that uses less energy to produce the same amount of light as an incandescent light buib is an example of energy efficiency. So, a householder's decision to place an incanadescent light bulb with compact fluorescent is an example of energy conservation. So, as individuals, every countries' householder choices and actions can result in a significant reduction in the amount of energy used in each sector of the economy.

So, I bring this interesting question: What factors can influence householder to choose to do any efficient energy consumption or useful behaviors at homes? I believe every countries' householders will have their different living attitudes and their living attitudes can influence their behaviors or activities to choose hoe to use energy at home. I shall indicate some countries' householders' living attitudes to explain the factors can influence them to use energy efficiency at homes as below:

- Is the low income and rising price of modern fuels both factors best to influence Nigeria householders choose to use energy efficiently?

Firstly, for Nigeria householders energy consumption habit at homes example, it is richly with natural resources, modern energy resources which provide many householders with biomass (mostly firewood) and some other householders modern energy sources, such as kevosene, liquefied, petroleum, gas and electricity for their use. So, it is one country which can manufacture to provide energy for itself to use. It doesn't need to depend on other countries to import any kinds of energy to householders to buy to use at homes. But, it has social challenge, the poverty problem in Nigeria goes beyond low income, savings and growth rate, due to its low level of education, poor governamce, high level of unemployment factors influence.

It is important to know how Nigeria householders meet their basic energy needs between poverty and energy can bde described in terms of quality and quantity of energy used. Generally, most poor householders use biomass fuels because of affordability and they (householders) do not have energy equipment (such as, gas cookers, electric cookers etc.) . So, it seems Nigeria householders won't demand their living quality to be improved. It implies that they will use any kinds of energy efficiently at homes, e.g. gas, electricity, due to they find themselves in energy poverty. Although, this country has enough nature resources to manufacture energy to provide to householders to use, but due to many people are low income group, so they won't spend too much expenditure to buy much energy to use at homes. So, the rising prices of modern fuels, such as liquefied, petroleum , gas (LPG) and electricity and their erratic supply have made many householders revert to the use of traditional fuel, such as firewood and charcoal.

It brings this questions: Is the low income and rising price of modern fuels both factors best to influence Nigeria householders choose to use energy efficiently?

The hypothes is predicated on the economic theory of consumer behavior. However, when income increases, householders not only consume more of the same goods, they also need higher quality . So, it applies economic theory to householder's energy consumption behavior at home. It explains why low living standards induce greater dependence on firewood and other biomass fuels owing to a combination of income and substitution effects, such as Nigeria low income household energy home users case. it explains why Nigeria householders can accept to use firewood and charaval traditional energy to replace liquefied, petroleum , gas (LPG) and electricity modern energy . So, economic theory explains the Nigeria household energy users why they can accept to use traditional energy to replace modern energy and their energy useful or consumption behaviors are efficient at homes. Although, Nigeria has enough natural resource to manufacture modern energy to supply to householders to use at homes. But, due to these modern energy products prices are raised to the price level of householders who can not accept. it causes to Nigeria householders only choose to buy the cheap biomass, firewoods to replace high price of modern energy products to use at home often. So, they can accept their quality of living to be fallen down. So, expensive modern energy product price is one factor to influence some countries' householders to choose to buy cheap traditional poor quality of nature energy, e.g. firewood or biomass, to use at homes. Hence, they can raise energy efficiency to use when they choose to use traditional nature energy to replace modern nature energy at homes.

- Does season factor influence New Zealand householders' energy consumption behaviors at homes

Secondly, for New Zealand householders energy consumption habits at homes , for example, their living quality needs are general comfortable need feeling. Their countries' houses of space heating was found to average 34% of total housholder energy use. The relation to space heating includes low indirect temperature are associated with persistent under-heating , whether some space heating sources tend to be higher or lower in winter indoor temperature than others and winter indoor temperatures are compared to international benchmarks and established healthy temperature ranges. So, New Zealand occupant's perceptions of winter indoor temperature conditions are presented and explored in relation to heating patterns and household energy consumption. So, it seems that NZ winter temperature is low. Moreover, it will influence householders need to turn on heaters to keep more warmer feeling indoor. Then, they will use more electricity energy. In special, if the householders' houses spaces are large sizes . Hence, their heaters need long time to keep whole houses' areas or spaces or rooms temperature to be rised up in order to let they do not feel very cold in winter. So, NZ's winter extreme cold weather will influence householders' energy use or consumption to be increased in winter.

The electricity efficiency to every NZ householder is very high in winter to compare spring, summer, autumn seasons. Hence, if NZ electricity suppliers expected to forecast electricity consumption more accurate in NZ. In order to ease the life for both electric net designers and electricity suppliers, it was decided to find out, how the NZ weather conditions and every householder's house space size factors to influence the power consumption to NZ householders. If there is a clear trend observed , then this relation can be used for power consumption forecasts to NZ householders.

Why does NZ weather condition factor and householder's house space size factor can predict householders' electricity consumption at homes. Due to geographic location on the global the lowest south sets specific conditions for weather, such as NZ's south island geographic location is near to south ocean in our earth. It is a country where average annual temperatures are well between 10 degree to below 10 degree at NZ south island special geographic location to near to the sourth ocean in our earth at the same time.

However, large part of mankind is living in the conditions where there are four different seasons in NZ geographic location, dark winter, which is cold and snowy, spring with rising temperature and high precipitation, sunny , dry and rather hot summer, and windy and wet autumn. These conditions lead to different patterns in electric appliances use in NZ householders, in special, in NZ south island householders. If trends in electric energy use have substantial correlation with weather conditions, this can help NZ electric energy suppliers and producers to forecast electricity consumption and thus organize and manage production of electric energy.

Consequently, it will lead to much more stability in energy supply to NZ every householder. For example, when the NZ energy supplier gathers data concerns every householder's house space size data, e.g. the house has how many sleeping rooms, toilets, bath rooms, eating rooms and reading rooms number, even the house has how many family members are living in every NZ geographical location. Then if it can follow different location of NZ houses spaces sizes whether they are large or small space size as well as whethe every house has how many family members are living to evaluate whether how much electricity efficiency can satisfy their comfortable living needs in winter. Then, it can evaluate whether they will use how much electricity efficiency for their needs in different seasons. If in winter, many householders are living in the large space size house in the geographic location. Then, it is possible that the geographic location is householders will use much electricity efficiency and where geographic location hosueholders who will be possible to pay the most highest electricity fee to compare the other geographic location of small space size of house householders. Hence, weather factor is the most influential to change NZ householders ' electricity energy consumption behaviors at homes.

- Urbanization level and income per capita both tangible factors as well as temperature (weather variation factor) will have close relationship to influence China householder energy consumption or useful needs at home every day

For China householder energy consumption habit example, what factors can determine to impact this country's householders energy

useful behavior at homes? Can the impacts of these factors be quntified? What are China householder energy consumption trends and characteristics? I shall explan as below:
I believe the influential factors include these three aspects to China householder energy users: Income per capita, urbanization level an annual average temperature (weather). These factors will influence any China householder energy useful or consumption behavior at homes.
Temperature (weather variation factor) is intangible from eastern region to western region of Chin, variances largely depend upon economic level and the provincial level. So, some regions were warmer and cooler temperature will influence the regional China householder how to use electricity. In addition, th influence of urbanization level varies according to income level as well as the urbanization level has more significant impact on the structure and efficiency of China householder energy consumption thatn on its quantity. So, the urbanization level and income per capita both tangible factors will have close relationship to influence China householder energy consumption or useful needs at home every day. Moreover, these two tangible factors (urbanization level and income per capita both factors) have the more influential to impact China any one of household family energy consumption or useful habit to compare temperature factor at home. Because temperature can only influence than to choose to turn on heaters to keep more cooler in summer or turn on air conditions (fans) to keep more warmer in winter.
The electricity energy needs for these equopment tools which will be influenced less. Otherwise, the urbanization level and income per family householder how to choose to spend more or less electricity or gas etc. energy at homes. Because in behavioral economy view point, when individual householder has more income and the urban in the China geographic location is lising many high income and high household families memebrs to every house. Then, the urbanization household energy household enery useful or consumption level will be raised. Such as China household

electricity users case, e.g. large cities have many high income and many houses have more than four families members to live on one house together. Then, the electricity or gas energy efficiency will be influenced to rise. The city urbanization and per capita income level is high to these large cities have high to income population, who are living in these cities in China.

Moreover, the impact of lifestyle on energy use mainly reflects types and purposes of fuels are chosen by different China households factor which will influence the urbanization level of energy choice use. China is a country with typical binary economics and social diversity and these is significant difference in the consumption pattern between urban and rural regions. Urban residents consume high-quality energy, such as electricity, natural gas , heating power, solar energy and gasoline. For rural residents, usually use coal, and bismass energy because they are cheaper price energy products which requires much time and labor and are heavy indoor pollutants . The difference in energy consumption pattern between urban and rural China residents is closely related related to living of quality needs, building structure, e.g. steel or stone etc. different materials, manufacture, easily access clean and effective feels through the electric grid, natural gas network and district heating systems.

Therefore, it explains why urbanization level is as an integrated variable reflecting social progress situation to influence urban and rural regions, such as large cities , small cities and rural countryside regions' household energy consumption or useful behaviors which have differnet kinds of fuel useful demands and energy efficiencies qualify and quantity demand, or needs at homes. Consequently, it explains, urbanization level and income per captia level both factors are more influential to China household energy consumption at home to compare temperature (weather , seasonal) factor.

- Employment rates or gross domestic product macro economic variation factor, residential space size factor, and the government's implementation of energy labeling schemes provide significant impacts on Taiwan residential electricity consumption .

For Taiwan householder electricity consumption characteristics in the residential sector, which has different factors and pattern to compare China householder electricity householder electricity consumption habit at home. Although, they are the same Asia country. I shall explain these reasons as below:

For Taiwan electricity householder factors influence their energy useful or consumption behaviors at homes. The main factors can influence their electricity energy useful patterns include: employment rates or gross domestic product macro economic variation factor, residential space size factor, and the government's implementation of energy labeling schemes provide significant impacts on Taiwan residential electricity consumption . However, the impacts of electricity raising price and the energy supply reducing shortage efficiency standards do not significant to influence the Taiwan residential electricity consumption behavior at sources.

It means that it won't influence Taiwan householders to use electricity or gas or any kinds of energy number to be reduced, even the Taiwan government energy suppliers sudden raise, any kinds of energy price and reduce to supply energy to satisfy Taiwan householders daily essential needs at homes.

In fact, Taiwan had improved gross domestic product (GDP) and it had raised employment rates recently. So, many Taiwanese has jobs to work, due to Taiwan economy had improved to be better. So, growth had also raised. The economy improvement causes many Taiwanese had enough jobs to work, due to new businesses are set up. Many consumers excit any kinds of businesses are invested to Taiwan from overseas or local investors. So, consumption is grown, the electricity consuming applicances are selected, as the household consumer focus grousp number if also influenced to be increased. So, Taiwan economy had improved to be better, it will encourage many electricity consuming applicances products are encouraged to excited to be selected to seel in Taiwan. Due to many different kinds of electricity consuming appliances are supplied to attract

Taiwanese to choose to buy to bring to their homes for cooking, boiling water, or keeping rooms to be cooler or warmer temperature confortable feeling intention in winter or summer seasons. So, these electricity consuming appliances, e.g. rice cookers, heaters, air conditions, fans, bathing gas heaters etc. different home electricity consuming appliances will be increased to supply to satisfy Taiwan householders' needs. When they decide to buy any news electricity consuming applicances to bring to homes to use.

● Environment scientists' education message how to influence Greece householders home energy consumption behaviors from primary energy to change secondary energy

Finally , I shall indicate Greece, this western which will influence this country's householders have desires to do household energy conservation patterns or conservation energy consumption behaviors or energy conservation activities at homes. I shall explain the social economic variable, such as consumers' income and family size variation factor which can influence the different Greece family household members differences towards energy conservation preferences. IN addition, the variable, such as environmental information feedback and consciousness of energy problems are characteristics of the energy saver consumer.

Why and how can environmental pollution , environmental protection, energy conservation information message can influence Greece householders to choose to do energy use consumption conservation or less energy useful behaviors at homes. It is one interesting energy efficient use behaviors , due to Greece householders are influenced by energy conservation or environmental protection message.

In fact, scientists agree overconsumption of natural resources is a major threat to oue lives in earth. Environmental problems like greenhouse effect, ozone layer depletion, and acid rain effect are not any more problems of a specific region or environmental problem. Also, economic theory is indicated that in order to gain comfort and time households are becoming excessive energy users, neglecting the environmental impact of their choices.

Environment scientists bring these environment pollution message to influence Greeks (Greece householders) to change their energy consumption behaviors at homes. The environment scientists' message indicate that we are facing global warmth and natural resource and energy shortage challenges. Due to our Earth have limited natural resource numbers to supply to us to manufacture energy, but global population has been increasing every year. Thus, it is possible that we have energy shortage crisis. Also, manufactures are spending too much energy to waste to manufacture any products, the energy will cause air or water pollution in manufacturing process or drivers are driving their vehicles to pollute air on the roads.

Hence, environment scientists' message influence Greece householders began to consider these questions concern to reduce fossil fuel energy. Why do we need to Safety in using fuel and handle gas leaks? Why do we feel town gas smell? How is electricity located at electric station far away from town area? How to solve problems caused by the use of fossil fuels? How to reduce the use of fossil fuels?

Greece householders consider to solve the problems, the best way is to reduce thir used of fossil fuel. This helps prevent fossil fuels form being used up too quickly. Also, it helps them to reduce environmental problems because fewer pollutants are given out when less fossil fuels are used. Can human help to reduce the use of fossil fuels? Fossil fuels are mainly in power station. Although they use some fossil fuels for our gas cooker and car, it won't make much difference if I use less. Fossil fuel is not used renew primary energy. Most of energy Greece householders use come from fossil fuels, for example, the electricity we use is generated in power stations by burning fossil fuels. The buses they ride use diesel oil. Therefore, they can help reduce the use of fossil fuels by saving energy in Greece daily lives.

The actions that Greece householders can take such as: setting the air-conditioner to a higher temperature, walking instead of using lift, taking a short shower instead of a bath. This reduces the use of the hot water and thus the energy needed to heat the water. Thus, many people can help a lot to reduce our use of fossil fuels to avoid fossil fuel shortage risk occurrence.

Greeks (Greece householders) had been beginning to conern that they will face energy shortage challenge if they can not adopt more energy conservation actions. Because the Greece government began to bring negative environmental pollution and energy shortage challenge message if they often waste to use any kinds of energy, e.g. electricity , gas excessive number efficiency at homes. Then, they will be possible to fac energy shortage and environmental pollution challenge to their country in future one day. So, this energy shortage and environment pollution message has bring predictive negative worries to influence many Greece householder energy home users choose to reduce to avoid the waste of any kinds of energy use at homes.

So, their reducing energy use actions that had encouraged them to cause habits to avoid to waste excess energy to do any non essential electric appliances useful or consumption activities at homes often. Moreover, the environment protection and energy conservation message has changed many Greece householder to make decision and activities to change their lifestyle to b low living quality from high living quality. So, the environment protection and energy conservation message factor has much influential to change Greece household energy users' daily energy conservation or less energy use consumption activities at homes.

Greeks feel greenhouse energy can be environmental protection enegy. A greenhouse can trap heat in the sunlight and keeps the air inside the greenhouse warm enough for plants to grow. The glass roof and walls of a greenhouse let in sunlight but prevent heat from escape, this makes the greenhouse warm inside. Similarly, some gases in the Earth's atmosphere can trap heat from the sun and keep the Earth warm. This is called the greenhouse effect. The gases energy that can trap heat from the sun are called greenhouse gases. It is future one kind of potential primary energy to reduce environmental pollution new energy products for human

consuming. So, environmental protection message influence them to consume greenhouse enegy at homes.
So, environment scientists' environment pollution message had influence Greece householders concern to apply seconday energy (environment protection) to replace electricity energy to use at home. They will change energy to use at home. The scientists' messages have more influential Greece householders energy change consumption behaviors at homes. The messages are as below:
There are different forms of energy, e.g. light, heat, sound, wind, water, electrical kinetic, chemical and potential energy. Some form energy is primary energy and it can not renew to use, e.g. light, sound, wind, water, fossil fuel etc. Some form energy is secondary energy and it can renew to use in possible, e.g. nuclear, electric charge battery etc. Why does human need to concern how to manufacture secondary energy? Because it is possible that our natural resource will be consumed all, thus we will face primary energy shortage risk. If human can invent any new form of man-made secondary energy to renew to use in order to avoid primary energy shortage to supply to use to use, then human won't only depend on our Earth natural resource energy supply numbers. We can invent any new secondary energy to renew to use again either replaces primary energy or instead of primary energy limit number supply.

What is energy change? For television energy change power case. Firstly, electrical energy changes to television power to be used by television itself, then it changes to light power, next it changes to light power. How to choose fuel form to use? Due to energy can change to different form of powers to supply different form of power advantages to supply to human to use, so it is possible that we can also invent any secondary man made renew used energy to change different form powers to supply us to use, e.g. nuclear energy changes to light or sound or heat form of powers ; electrical charge batteries changes to light or sound or heat form powers to satisfy our daily life needs.

The environment scientists' energy consumption education influence Greece householders concern how to change to use secondary energy to replace primary energy at homes as below:

For primary natural resource fuel energy example, different fuel has different feature, e.g. easy to burn, safe to use, gives out a lot of energy, inexpensive, produces little air pollution, easy to transport and store. How can we use in different channels, such as heating food, hot pat, driving vehicles.

For example, although coal is not expensive to cause electricity energy for past transportation tool, e.g. traditional coal energy train or our daily home cooking, but it has negative influence to environment air pollution. Hence, we ought to follow the primary natural resource energy's feature to decide how to apply what aspects of our life needs.

For example, if the country's people hope to reduce pollution when who use any kind of energy, e.g. US , Europe energy markets. The energy entrepreneur ought concentrate on manufacturing the kind of energy which can reduce environment pollution to be the least level to supply the country people to use, e.g. electric charge battery supplies to these countries' drivers to drive their vehicles on the roads, wind energy or water energy to manufacture electricity power supply to reduce air or water pollution ; or if the country people hope to buy the inexpensive energy to use, even the energy's quality and performance is worse, e.g. China, India, Hong Kong markets. The energy entrepreneur ought concentrate on manufacturing the lowest cost and enough supply of natural resource to manufacture the kind of energy to sell cheap price to these countries to use, e.g. China, Africa can accept to use e.g. gas, coal, fuel energy to use to compare developed countries people, e.g. UK, US; or if the countries people who hope to use energy which can easy to transport and store, e.g. light coal. The energy entrepreneur can choose to concentrate on manufacturing much coal to supply to the countries people to use, e.g. China, Arica Thus, to choose to manufacture which kinds of energy supply to the countries market people to use, the energy entrepreneur how decides to manufacture which kind of energy, it depends on which kinds of fuel advantages of the countries people most concerning.

What is energy meaning? It is defined a dynamic quality, it is a fundamental entity of nature that is transferred between parts of a system in the production of physical change within the system, and it is usually regarded as the capacity for doing work, and it is usable power (such as heat or electricity) or the resources for producing such power.

Why does secondary energy own investment worth? Because the different forms of primary natural resource energy will have supply shortage crisis, such as natural resources coal, gas, solar, wind, water, geothermal, biomass(organic material) etc. However, human can attempt to explore any undiscovered Earth or Space resource to manufacture any kinds of secondary energies, e.g. nuclear energy, electric recharge battery energy to supply to electric vehicle or space robots transportation tools to use or satisfy our daily life needs in future one day. So any kind of undiscovered secondary man-made renewed used energy resources have potential commercial worth to any energy entrepreneurs, it is possible that they can replace traditional primary energy to supply to human to use for our different aspects of life needs. In the future, the secondary energy demand will increase, when primary energy supply number has decreased form natural exploration. So, it will cause the effect of any demand of secondary energy product to be raised and prices to be increased in possible. Due to global population has been growing up, considerably China and India both countries populations have been increasing rapidly. Scientists predict there are more than 1.2 billion people worldwide will lack access to electricity, and more than 2.5 billion still use wood, charcoal to cook and heat in the future when primary energy has no enough number to supply to us to use. Hence, the fact that demand is this much greater than supply to make energy a prime market for further growth.

Although, secondary energy will have much investment worth, but energy like all other investments will carry risks. The internal and external risk factors include such as: policy is always changing to prohibit which do energy trading more easily between the energy exporting and importing countries, the secondary energy manufacturer itself own abilities to invent and to manufacture any kinds of secondary energy, improved technology can quickly make an technology obsolete, geopolitical rifts can happen overnight, the country's energy consumer (user)'s preferable choice to use which either kinds of secondary energy or secondary energy. So, it seems that (man-made) renewed used secondary energy industry can provide above-average returns, but it can also bring high risk commercial investment.

Traditionally, energy supply companies will apply those methods to operate energy providing businesses. For Shell,. Exxon examples, which had own gas stations, explore and drill for gas on their own. Other companies specialize in a part of the energy market, e.g. leasing oil rigs for example, or operating a pipeline. Energy supplying companies can choose to manufacture any kinds of energy to supply, e.g. trade oil, gas, coal, uranium, electricity etc. Any energy price and supply is demanded on the countries energy users' which kinds of energy most choice need or certain energy commodities to be chose to use popularly. For example, if US most people prefer to use secondary man-made renew used energy more than primary energy. Then, US energy manufacturers ought concentrate on manufacturing much different kinds of secondary man-made renew used energy to prepare to supply to its domestic US market in order to raise secondary energy price to sell in its country. So, the energy manufacturer's energy manufacturing choice, it is depend on which the country's people prefer to use which kinds of energy for their daily life needs.

However, scientists predict secondary energy market will have large market share, due to primary energy will have shortage to explore to supply in our earth and future energy consumers(users) prefer to choose to use more efficiency, less energy consumption, none environment pollution cause, cost effectiveness, renew to use of any kinds of energy. For example, the electricity recharge battery secondary man-made renew used energy is one kind of reducing air pollution power to push any electric battery vehicles to be driven to compare gas energy during drivers are driving their cars on the roads. They can reduce noise and air pollution and drivers can drive safely, who only need to buy one electric recharge battery to recharge in any electric recharge battery stations on streets when the electric recharge battery has no enough power to push their cars and they need to recharge their electric recharge battery drive when they had driven between one to two days. Due to primary energy, e.g. fuel , gas, the kinds of primary energies will have shortage to supply to global drivers to drive their traditional cars. Thus, the electric recharge battery or any undiscovered secondary energy will be future driving market needs. So, man-made renew used secondary energy, e.g. biofuel, hydro-electric, nuclear, will be one kind of efficient, clean, less pollution cause, cost-effective of energy to supply to our global vehicle market, even any other undiscovered new markets. Supposing they are popular to be used for electric vehicle market globally in future one day, then their prices will be decreased and constructed to average car requires up to 1,700 gallons of oil. Also supposing that making average computer requires more than ten times or weight to fossil fuels, every calories of food eaten in the US requires roughly then calories of fossil fuels. Hence, cheap energy will be one successful factor to influence future potential energy consumer (user) individual choice needs. Conversely, ion good economic times, people are more willing to travel, to buy products, and all of which success demand and low process for energy.

In the future, secondary energy will be the best choice to food production market. The modern food production system is essentially a success of changing fossil fuels into food. So, raising energy prices are almost higher food costs and even shortage for fossil fuels energy. If one day, one kind of discovered secondary man-made renew used energy can supply to any restaurants or homes to be used to cook at the cheap price, then the profit is very high for this kind of food production energy. Thus, future food production secondary energy consumption market is large and because the primary energy inputs for agriculture are higher than the energy outputs of the food. However, future secondary man-made renew used energy for food production system is only one part of whole energy consumer in food industry. The food production is related to whole food consumption market which includes: household cooking energy market, agriculture or vegetable, rice, fruit etc. foods farming machines energy market, food manufacturing factories market, food machine package market, transportation food delivery market, supermarket or fruit/ food sale stores market. They must need any energy inputs to achieve the food production or food transportation or warehouse / stores electricity supply or cooking energy needs. Hence, these food suppliers relate to any whole food factory manufacturers, food retailers, food wholesalers, farmers and home/restaurant cookers, all of them must need to use energy to carry on their food producing or food cooking or food transportation activities every day in overall food industry. Thus, it seems that undiscovered any second energy demand will be increased, when the primary energy supply number is decreasing. Also, when people can accept to use secondary energy to replace primary energy to be used for any cooking, transporting food, manufacturing food, food retail stores or warehouse food delivery energy need activities. Then, the secondary energy price will be fall down to attract many food energy consumers.

Nowadays, the food industry energy may includes primary nature resource gas energy or electricity energy for house house families or restaurants cooking needs, food delivering lorry drivers driving needs usually. If future second man made renew used energy is invented successful popular to be used, e.g. hydrogen, electric recharged battery energy for electric vehicles or restaurant/ home families cooking needs or food factories machine maufacturing energy needs. Then, the seconday energy will have possible to replace primary energy to be food industry energy market.

Wiley, composition services graphics indicated that global primary energy consumption had been increasing 30 billion tons from 1830 year to 510 billion tons in 2010 year as well as global population size had been increasing from 70 billion 1830 yeat to 510 billion in 2010 year. Thus, it seems that global primary energy consumption will be needed largely after 2010 year. If future global nature resource primary energy is explored full number and it had not enough energy number to supply global human to use. Then, it will being many people feel uncomfortable and inconvenient,e.g. Some countries won't have enough energy to supply transportion tools to be driven, some homes and restaurants won't have enough energy to supply to cook to eat or to provide restaurant clients to eat etc. daily activies, due to human's much activities which are needs energy supply. Thus, it seems that global primary energy comsumption will be needed largely after 2010 year.

Wiley, composition services graphics also explianed that why the primary energy consumption demand can be needed to achieve the same level to the global population size increasing in 2010 year. The graph showed these reasons why cause the same level of global population size and global primary energy consumpion demand which may include: The graph showed that after a nation is developed, its per-person energy use hegins to level off. In North Ameruca and Europe, where energy demand has remained flat, or fallen dightly, in each of the past few years. But the 1.3 billion people on those two continents are far outweighted by the 5 billion people in Asia and Africa, e.g. Chinese and Indian. who currently have more energy need to comapre average per man to North America and Europe per man, ensuring that overall energy demand will rise for years to come.

Wiley, composition services graphics also predicted that the growth in primary energy demand. China will have 4,500 million tons in 2035 year. India will have 3,000 million tons in 2035 year. Other developing Asia will have 2,000 million tons in 2035 year. Russia will have 1,500 million tons in 2035, Middle East will have 1,300 million tons in 2035, other rest of world will have 1,000 million tons in 2035. Hence, it implied that China will be the largest primary energy need country in the future.

China will be future the primary potential energy consumer market. The primary energy includes water, coal, wind, fossil oil, gas ,solar, geothermal energy, biomass (organiz material) etc. different natural resource primary energy. Otherwise, US, UK, Europe will be secondary energy potential need market. For example, electrical recharge battery energy will be raised demand to supply to any future new design electrical charge battery vehicles in US, Europe, UK markets.

Due to US, Europe, UK people concern environment protection, so they will invent many electric charge battery vehicles to consume electrical charge battery to replace polluted gas energy to avoid air pollution when the drivers are driving cars on themselve countries' roads. For example, second man-made renew used nuclear energy can be applied to rockets to pusch them to leave our earth to fly to other space far away and consuming nuclear energy will be cost efficient, and nuclear energy saving will be more when nuclear to spend long time to be used in any long time space journey. Hence, nuclear energy and electric charge battery secondary energy will be popular to be applied to vehicles and rockets energy needs in US, Europe, potential marketss, even our daily energy needs in global second energy market.

Who are your energy business's competitors (peers)? How do they compare? How have your energy business company performed cyclically? How to choose to manufacture to sell which kinds of primary or secondary energy product(s), either manufactures only primary energy product(s) or manufactures only secondary energy products or both? Which countries do you plan to sell your energy product?

Illustration by Wilsey, composition services graphiss showed that these natural resources to energy product the world's electricity percentage, such as below:

41% of coal, 5% of oil, 21% of gas, 13% of nuclear, 16% of Hydro, 3% other renewable secondary man-made energy.

Hence, coal will be future the major natural resource to produce electricity. The energy entrepreneur ought attempt to explore any coal resources, when who choose to supply electricity power to consumers for future energy consumption country markets.

Wiley, composition services also predicted that the expectation is that North America coal will supply the expectation is that North America coal will supply Asian demand, Us export terminals have a total capacity of 173 million tommes output. China will drive 16% of the nations total output. China will drive the sea-born demand for coal over for the forcessable future. Chinese energy consumption will grow more than 12 % between 1980 and 2009 years. Though, China heads global demand, India is growing faster in terms of coal imports. Much of the global coal demand will be supplied by Indonesia and Australia. Colombia, Russia, South Africa and Mongolia are also players in global export coal energy resources.

Hence, environment scientists' education messages influence Greece householders believe that secondary energy will be one kind of new energy product to replace traditional primary energy product for human energy consumption market global needs. Hence, it is right time any energy entrepreneur needs to research how to explore any undiscovered man-made renew used secondary energy products to avoid primary energy shortage crisis occurrence. Greece householders will be the highest population number to choose secondary energy to replace primary energy to use at homes. it means that environment scientists had changed Greece householders' energy consumption behaviors at homes.

In conclusion, different countries will have different factors influence how the country's householders energy consumption behavioral changes. Hence, it seems that any country's householders' energy use of consumption behaviors will be possible influenced by extermal environment factors influence. Also, every country's energy providers can attempt to find whether the country has what kinds of unique factors to influence its householders' energy consumption efficiency to increase or decrease in order to find the methods to solve the energy efficiency demand reducing challenges successfully.

AI technology how impacts to avoid food shortage

Why environmental factor can influence food shortage

Can artifical intelligence measure which the lowest environment pollution level influences food consumers' good wastage behavior rises ? What are objective indicators of standard of living and quality of life? Objective circumstances refer to the economic and material conditions which are important aspects of the standard of living and quality of life. In the assessment, eight different indicators were used: CPI, GDP per capita, shopping basket, household's expenditures, GFIC basket, poverty rate, income inequality and HDI. However, these indicators is one number measure. It can't measure anyone's psychological feeling, such as health, safe emotion. The challenge concerns whether environmental pollution factor, such as air pollution, water pollution can cause human's health to be poor, even goes down human's quality of life and economy loss. I shall indicate some evidences to give reasons to support my conclusion why I believe that environment pollution is a factor to cause human quality of life to be poor , even it can also cause economy will encounter loss too.

In general, measure of quality of life need include human's psychological feeling indicator. I shall indicate, Hong Kong, China countries air and water environmental pollution challenges how to influence these two countries' people quality of life to be poor, even, it will cause their economy loss. Nowadays, China and Hong Kong and India and Afria are encountering health problems arising from damage to lungs, heart and blood vessels. Hong Kong and India and Afria and China e.g. Shanghai city pollution is a significant cause of premature death from cardiopulmonary disorders. Present level of pollution cause injury to the immature developing lings of children and adolescents. This damage will lead to life-long health problems in many and a reduction in life-expectancy. Although, there is no evidence from analyses of trends in pollutants that pollution measures in recent years have reduced pollutant concentrations in a way which will benefits public health.

There are clear indicators that for some pollutants. The problem is worsening. In fact, air and water pollution is Hong Kong and China and Afria etc. developing countries' the biggest cause of social and environmental injustice. It harms not only citizens today, but because its transquenerational effects on the urborn and youngest members of the society, it will cause its will health effects well into the later years of this century, even environmental pollution challenge will cause these countries will encounter economy loss.

Human activities have created forms of air and water pollution, such as gases from fuels, uncontrolled emissons from fossil fuels and other chemical sources have long been recognized as a cause of ill health and premature death. For example, in December, 1930 year, a dense fog affected the Meuse Valley in Belgium. Beginning on December, 3 date, the fog intensified over three days and was associated with laryngeal symptoms, chest pain, coughing, and breathlessness. Some patients showed signs of pulmonary oedema. Overall 60 deaths were attributed to the episode. After a long investigation, the cause was considered to be emissions from high sulphur fuels, including suplhur dioxide and sulphuric acid.

What is the current threat to health? the migigration of air polluton following the introduction of clear air has been followed by a period of unprecedented economic development creating new forms of pollution from the combustion of fossil fuels. For example, in constrast to the relatively large tar laden particulates from burning dirty coal which caused episodes like the London city, UK. Smog , traffic pollution now genertes fine with a different size and composition and gases,such as which may cause injury to the respiratory system and the effects of other pollutants. Such as particulates and drive the formation of the secondary pollutant ozone. The effects of pollution will therefore to some extent reflect genetic, environmental lifestyle and behavioral factors to develop these distance in a population together with the existing prevalence of diseases which may be polluted.

Hence, living in polluted urban environments is associated with increased levels of biological markers of inflammation compared with residence in a clean air environment. The damage is caused by air pollution manifests itself through a variety of common and recognized health problems, such as upper complaints heart and lung disease. Because of this, we can use statistical methods as well as clinical studies to detect the signal of changes in health problems and increased health care demands in the population. However, doctors had proved air or water pollution can cause these both curdiovscular or respiratory disease indirectly. Curdiovscular disease includes formation of arterial plaques, coronary artery, heart attacks, irregular heart rhythm, loss of heart rate variability, high blood pressure, stroke etc. disease. Respiratory disease includes inflammation of nasal, throat and tracheal airways with acute, lower respiratory tract inflammation and infection causing bronchitis, reduction long growth and function in young people. So, it seems environmental pollution can influence quality of life to human as well as environmental pollution and illness and poor health problem has close relationship.

On the other side, envionmental pollution can bring health risk, over it will influence social inequalities. Some researchers had found that the evidence has been compiled for six envionmental health challenges, such as air quality, housing and residential location, unintentional injuries in children, work related health risks, waste management and climate change. It seems human need to concern air and drinking water quality, waste management and climate change how to influence our environmental pollution challenge. Although, the evidence base on social inequalities and environmental risk is fragmented and data are often available for few countries only, it indicates that inequalities are a major challenge for environmental health policies. Irrespective of development status, environmental inequalities can be found in any country for which data are available. The valid for the exposure to environmental risk factor is also unequally distributed, and this unequal distribution is often related to social characteristics, such as income, social status, employment and education, even environment risk factor can influence human's quality of life.

(i) How environmental risk factor can influence different groups
However, human need to concern how environmental risk factor can influence inequally health outcomes to different groups. Such as, the first group is social determinants affect the environmental conditions of an individual and may contribute to the fact that specific individuals or population groups more often experience loss adequate or potentially harmful environmental conditions. The second group is the affected population groups could still be more exposed through e.g. the mechanism of education and health behavior. The third group is given socially disadvantaged groups could show more severe health effects of the social disadvantage is associated. The final group is social determinants affect health (what remains unclear is the relative importance of socially determined exposure to environmental risk factors). Thus, human need to concern our behavior can lead environmental pollution to influence poor health to alive. Even, we can not neglect how to protect our natural environment to be clean issue. Due to environmental unhealth poor issue can lead our bodies to be unhealth and to be ill and we have no health to work to influence our job inefficiency and low productivity if we often need to see doctor to raise workload to my staffs often. Then, our employers will be probable to dismiss and many unhealth employee will lose jobs and unemployment ratio will raise and DP will reduce, it will influence our economic growth . Hence, we can not neglect environmental justice and environmental inequity issue, e.g. indoor air pollution and occupational or exposure to environmental tobacco smoke pollution exposure to high traffic roads or to industrial plants pollution in our society.

(ii) How Afria country environmental pollution influences
Surprisingly, most of above countries , among of them, although Africa is a green and natural environmental country, but Africa has encounted poor natural environmental quality to influence it

has poor quality of life to its citizen and poor economy growth to its society both. Why does Africa encounter this natural environmental pollution challenge? Afican have now two potential sources of pollution: consumption and production . This looks reasonable to Africa, since maintenance is completely dedicated to improving the environment, when production generates pollution only as a " by product". Capital implies the possibility of a country being trappical in an economent poverty trap by both a bad environment and low longevity. Some countries (or regions) may even experience other time, both environmental degradation and decay in expectancy. The fact that, in some cases, environmental degradation doesn't imply lower longevity may be due to the fact that economic growth might , at the same time, worsen environmental quality, but generate additional resources that can help increasing (or preserving) longevity. However, these is also evidence of countries where environmental degradation is associated with a reduction in life expectancy. It seems worsen environmental quality will influence any country's economic growth and poor quality of life both. For example, McMichael et al. (2004) identify 40 countries that experienced a loss in longevity between 1990 year and 2001 year (26 between 1980 year and 2001), they also support that the resulting world divergence in terms of life expectancy might be explained by ".... (the growing) health risks consequent on large-scale environmental changes is caused by human pressur, by both bad environment and low longevity, biodiversity and sustainable energy".

(iii) How human adult consumption and environmental quality influences future environment for human survival probability of life expectancy.

I shall assure human adult consumption and environmental quality has relationship to influence the future environment (green preferences) to provide human survival probability, it depends on inherited environmental quality. Thus, human will increase or decrease in the survival probability when we need a higher or lower life expectancy. In general, we depend on these environmental conditions to live, which include quality of water, air and soils etc. and resource availability, biodiversity, forestry, fisheries etc.

It is interesting to analyze different possible strategies to escape from the environmental poverty trap as well as factors that could push some economies back to a low equilibrium characterized. To research whether environment factor has relationship to influence human quality of life. We need to give idea of explaining whether environmental care has relationship to an uncertain lifetime. However, I suppose that an environmental kind of factor can be instead of being defined in terms of GDP per capita, capital accumulation etc. economic factors. Poverty is now related to environmental quality. It should be clear, however, I focus only on one specific mechanism lying behind environmental traps. Just as under development traps may be related to a wide variety of factors, ranging from financial to technological ones, including human capital accumulation and life expectancy. So, I should use this assumption to explain why it has relatively between environmental quality and life expectancy.

This " synthetic" indicator (YCELP, 2006) indicated environmental health is defined by child morality, indoor air polluton, drinking water, adequate sanitation and urban particulates and ecosystem vitality that includes factors like air quality, water and productive natural resources, A key ingredient of our setting is that survival until the last period is probabilistic and depends on the inherited quality of the environments. This survival probability affects the weight of the future environmental quality in human's utility function to achieve interest aim. Final stage, human will have optimal choices depend on life expectancy: in particular, a higher probability to be alive in the third period boosts investment in the environment and reduces consumption. In this case, a given country may be caught in a high morality/poo environment if low income is associated with a deteriorated environment.

John and Pecchenino (1994) were the first to introduce the possibility of multiple identifying, case for a poverty cause characteristic by poor economic performance and environmental degradation, however, life expectancy is assumed to be exogenous and plays no role in their model. Such as soils deterioration are the like, are all susceptible of increasing human morality (thus reducing longevity). So, the existence of both environmental performance and longevity, with countries being concentrated around two levels of environmental quality and life expectancy respectively. The two-way causes are between the environment and longevity. If the causal relationship between environmental quality and life expectancy involves the existence of an environmental poverty, characterized by both bad environmental conditions and short life expectancy.

Human life stage will encounter generations of three periods to get utility from consumption and environmental quality. During adulthood, when all relevant decisions are taken, adult can work and allocate their income between consumption and investment in environmental maintenance: consumption involves deterioration of the future quality of the environment (through pollution and/ or resource depletion) when maintenance helps to improve it. The dynamics of environmental quality may also be affected by external factors on more resourced communities. The most importance, unhealthy physical environments across the region adversely affect everyone, ever though who are likely to be most concentrated in more burdened community which also have less social power to change those environments.

Why life expectancy and the environment has close relationship to influence quality of life? Life expectancy and environmental quality dynamics are jointly determined. Human may invest in environmental quality, depending on how much , we expect to live. However, environmental conditions affects life expectancy. In particular, some countries may encounter in a low life expectancy / low environmental quality. This outcome is consistent with stylized facts relating life expectancy and environmental performance measures. Some expects to live longer, who would be willing to invest more in environmental quality, because who feel which have causal link between life expectancy and environmental quality. However, environmental quality is a very important factor affecting health and morbidity: air and water pollution, depletion of natural resources and quality of life.

(iv) Why social and physical environmental factors have close relationship to influence economic growth or
food shortage causing

I shall indicate reasons to explain why social and physical environmental factors have close relationship to influence economic growth, even human health of quality of life. The social and economic burdens of poor education, lack of affordable housing and less than self sufficient income affect, not just those individuals and families who have the fewest resources. The social gradient means that not only do whose in the bottom worse health outcomes to bottom of income group and the top income group whose will have poor quality of life influence. The higher rates of disease and disability and lesser productivity among many communities means a higher public and private burden of life years, particularly life expectancy once one reaches age 65. In recent decades, research and has increasingly shown how powerfully social and economic conditions determine population health and differences in health among subgroups, much more so than medical care. It seems that environmental factor can influence human's quality of life.

Los Angeles Country Department Of public Health (2016) indicated a country health rankings model, this department explained these three health factors can cause this health outcomes. These health factors include health behaviors (30%), it includes tobacco use, diet and exercise, alcohol use, unsafe sex; clinical care (20%), it includes access to care, quality of care; social and economic factors (40%), it includes education, employment, income, family and social support, community safety; physical environmental factor (10%) , includes natural environmental quality, built environmental quality. Then these factors can cause this health outcomes, such as morality (length of life):50% and morbidity (quality of life) :50%. SO, it implies that physical environmental factor can influence human's length of life. So, on our social environmental problems result is from a complex interplay of a number of forces. An individual's health –related behaviors , particularly diet, exercise and smoking, surrounding physical environment and health care (both access and quality) all contribute significantly to how long and how well human love. However , none of these factors is as important to population health as are the social and economic environments in which human live, learn, work and play. We refer to these factors can be as the social determinants of health to influence our quality of life. How do social determinants affect our quality of life? In the late 19th and early 20th centuries, public health concentrated particularly on the physical environment. Improvements in, for example, clean water supplies, healthier housing, sanitation, workplace safety and safe food lead to sharp increases in average life expectancy . Also our quality of life needed to be concentrated on expanded access to medical care, resulting in further expansion. So, the poverty tap is now characterized by those elements, such as low levels of : (i) environmental quality, (ii) life expectancy and (iii) human capital.

In fact, environmental degradation can have a significant impact on human health. De Hollander et. al (1999) & Melse & De Hollander (2001) showed that estimates of the share of environment, related human health loss are as high 5% for high income countries, 8% for middle income countries and 13% for low income countries. Air pollution and exposure to hazardous chemicals are important causes of the related burden of disease in countries. The transport and energy sectors are major contributors to air pollution, when important sources of chemical pollution are agriculture industry and waste disposal. Opportunities for reducing environment-related health risks are considerable. The benefits of many environment policies in terms of reduced health care costs and increased productivity significant exceed the costs of implementing those policies. So, the impact of environmental risk factors on health are extremely varied and complex. For example, the effects of environmental degradation on human health can range from death caused by cancer, due to air pollution to psychological problems resulting from noise. So it implies environmental factor can influence our quality of life in our societies. However, many factors can also influence human's health of a population, including diet, sanitation, socio-economic status, literacy and lifestyle.

De Hollander et. al. (1999) & Melse and De Hollander (2001) showed that total burden of disease, with estimated environment-related share expenditure, mid-1990 year. The average income group has 15 daily/1000 capita, the middle income group has 20 daily/1000 capita, the high income group has 10 daily/1000 capita. As regards both total burden of disease and the health conditions related to environmental; degradation. The result indicates the environment –related share of the burden of disease is greatly dependent on income, with higher-environmental shares generally occurring in lower-income countries.

On the one hand, it seems the large environmental share of health problems is primarily, due to factors related to poverty, such as limited to access to proper food, housing, health care and drinking water. Environmental determinants of human health in developing or developed countries are related. On the other hand, those to the exposure to air pollutants (particularly in urban areas and chemicals in the environment than to poor living conditions. Also sources of human exposure to chemicals are many and varied. Chemicals can reach the environments, for example, through emissions from industries, anti-fouling paints on marine vessels, pesticides in agriculture, waste incineration and leakage from waste disposal sites. When emissions of chemicals from industries and other point sources of pollution have lead to poor quality of life, source of chemical exposure. Intensive agricultural production uses chemicals in pesticides and fertilizer and in feed additives and medication for livestock. Residues remain in fruit, grains, vegetables, meats and daily products, all of which can reach the consumer.

Other sources of chemicals in food include bio-accumulative chemicals in the environment, such as heavy metals and persistent organic pollutants, which can be found in fish, meat and dairy products. So, environment pollution can influence human need to eat bad or unhealthy food to cause we have poor quality of life to live, such as the high income group or middle income group or low income group of families in our societies fairly. Other human health risks that have recently received considerable attention include unsafe livestock feeding practices through which toxins reach the food chain unintentionally. Dioxins that have accidentally contaminated poultry feeds that contain diseased animal remains can cause the so-called " mad cow disease" in livestock which has been linked to a new form of disease. The effects on health from exposure to chemicals and air pollutants vary from allergies to cancer. Although, the link between exposure and disease is often not clear, Even at low exposure levels, urban are pollutants can cause, asthma, allergies, respiratory diseases and cardiovascular disease if the exposure is continuous or long term. Heavy metals have been shown to cause neurological disorders and various cancers. In addition to , physical diseases, environmental contamination can also cause psychological problems. Noise, one of the determinants of the quality of urban life can have an impact on human health, decreasing the quality of life and potentially contributing to depression.

For Ireland, UK country example, this country politicians and policy makers believe the role of environment can be used to measure quality of life, concerning on either in its own right or relative to economic and social aspects of quality of life. Agreement on what measures quality of life and how it can be measured by the role of environment, not just in Ireland, but everywhere. The conventional approach is used for policy has been to use measure of gross domestic product(GDP) or regional valued added. However, it is acknowledged that such conventional economic measures have only a partial relationship with societal wellbeing. To the extent that economic measures are related to public products and consumption, there are also pressing issues in relation to public products and the sustainability of economic growth. However, the role of environmental factor can influence resource use and human's behavioral consumption.

Aspects to quality of life other than income include the environment, freedom, health, working condition, leisure, social and family relationship. Economists don't deny that these factors do play a role in quality of life. However, environment factor can be one role to influence other factors to influence our quality of life to be good or bad effect. For example, locations which might be desirable as paces to life (in terms of income earning opportunities or other factors) were also likely to have higher costs of living, particularly with regard to house prices or health or unhealthy air/ water pollution of environment situation of the place to provide human to live. Alternatively, social indicators are based on normative ideals of literacy, low rates of premature mortality or a quality environment. Other measurement of people's personal evaluation of their quality of life, much depends on personal expectations and experience.

(v) How environmental factor can influence any country's house price and food shortage causing

I shall indicate that why environmental factor will influence any country's house price. For Ireland example, citizen average incomes and higher in the east of the country, house prices are lower in the west, who are also more able to afford a property of choices. There are more opportunities to purchase houses, where people own their own houses, who are more likely to have benefits from an appreciation is its value and to consequently perceive a higher degree of health. Generally, levels of property appreciation have been higher in the east. Unfortunately, young people and the more economically active segment of the population are more likely to be faced with rising entry level house prices and the prospect of large borrowings. So, the quality of life, such as education, crime and access to healthcare and living environment are not uniformly better in the west or the east regions. Indeed, many measures of social disadvantage are at their worst in the west regions. Some indicators of environmental quality are , indeed better in the west regions, but there are others, such as drinking-water quality or recreational access that are often worse.

Comparisons can often be reduced to an urban-rural dimension rather than a regional one. Factors such as incomes, house prices, crime levels, air pollution and congestion are all likely to be higher in urban areas in Ireland city, UK country. Why environment and housing price has relationship in Ireland to influence quality of life to its citizen? If Ireland's regional development policy is successful , it will bring with it greater competition in the housing market and greater pressures on the environment in Ireland. Because the forest will be decreased to build house, the natural environment will become wood and steel and stone of housing built environment. In fact, it appears that there is a fair of amount of agreement on the relative rating of factors influencing quality of life. Ability to own one's home and security of income were needed, but respondents also placed almost equal important on clean air and drinking water, low crime were differences. The Ireland's rural respondents appeared to place a slightly greater emphasis on key natural environmental attributes, when urban residents valued absolute incomes and social or leisure activity rather more.

In this respect, the analysis identifies three components to Ireland people of quality of life, each of which was evident in all three locations. There components can be broadly described as domestic security, social/leisure and aspects of the planned environment. The first of these includes indicators, such as security of income, absolute income, house ownership and low crime. As this component includes air and drinking-water quality, it suggests that these indicators may be associated with personal health and well-being. When the planned environment component includes those attributes that affect quality of life over which the authorities have a direct influence, for instance, a clean environment, traffic and reducing vehicle numbers on the roads in busy time.

(vi) How environmental pollution can influence social welfare and cause food shortage

Environmental quality has an undefined impact on quality of life and various indicators are used to show regional variations in aspects, such as water quality . There are many measures of environmental quality , but is only for quality of life. Moreover, the measurement of societal welfare is important. Societal welfare is not simply , the sum of the parts, but varies depending on the individual in which people find themselves at any time in their life. In principle, it should be possible to apply weights to each element of societal welfare, but as preferences for each of these vary within the population. In the absence of a method with which everybody is satisfied, GNP and GDP are typically the most popular used measures for quality of life or standard of life. But, these are problems with the data itself to measure quality of life because quality of life is feeling or satisfaction of level to the country's citizen and it can not be seen by numbers or statistic method. For example, GDP ignores household production, such as the effort that goes into the rearing of children, the benefits that this provides for society and the public expenditure that is avoided. Neither are costs treated equally with the benefits. GDP counts all economical activities irrespective on pollution appears to increase. GDP even through it is a degree of double counting. Otherwise, environmental products are good to be measured to quality of life. For example, many environmental products are unpriced. Consequently, environmental products that people value, or which are critical to the sustainability of development, are abused or depleted because of their public products have good characteristics and the absence of a market price signal.

Environmental economists try to work within the economic model to measure quality of life. Rather than questioning the link between utility and consumption or choice, the preferred approach is to add an element into the utility function that represents the value of environmental products or the stock of natural capital. By one means or another , the preservation value of these environmental products is estimated in terms of willingness to pay to protect the environment or as willingness to forego other products in return. It seems the quality of people's environment can be represented by objective indicators. At another, their interpretation will vary and can be represented by subjective indicators.

Objective indicators come in two forms: (i) economic indicators and (ii) social indicators. The former depends on an ability to select the products and services that are desires, in other words, the satisfaction of preferences . The economic argument is that people select the best quality of life, who can obtain commensurate with their resources and personal desires. By comparison, social indicators are based on normative ideals on what could be considered the food life. For example, would be infant morality, literacy, crime rates and social indicators are objective measures. Both have guided, much of the research on quality of life, particularly concerning with the urban environment. Quality of life can include natural a significant influence on local quality of life, for instance, natural beauty spots used for recreation.

Whether environmental factor is the main factor
to influence food consumers to waste food

Can artificial intelligence find what the main environmental factor influences food consumers food wasting behavior rising ? For environmental quality concept, it concerns with health, safety, wellbeing, residential satisfaction and the physical sustainability can be considered to result from an when live ability can be considered to represent the interaction between the physical and the social domains. As with expenditure on the environment, investment in social capital contributes to quality of life. However, the benefits will again vary amongst individuals, depending largely on the security of their individual circumstance. As with the environment, the government can certainly adopt strategies that provide for public security by taking measures to reduce crime, a measure likely to be appreciated by everybody (except criminal) , at least to one degree or another. In other necessary to enhance social interaction, namely community centers or sports facilities. Furthermore, the creation of social capital has an statement which responds to general social trends to raise Ireland citizen's quality of life.

I shall indicate Ireland to explain whether environmental factor is the main factor to influence our quality of life and economic growth as well as to cause food wastage. Is environmental quality higher in the Ireland west regions? And if so, does this compensate for lower incomes in these regions? Is it bad that rural areas are characterized by higher costs of living in areas other than housing by environmental factor? In fact, in Ireland , UK country, population increase has a direct impact on the environment by placing demands on local natural resources, particularly open space and water. It also leads to a sense of crowding that reduces the utility associated with access to the environment. How can environment factor influence economy growth in Ireland? In Ireland, agriculture has gone through a period of significant change that has been accelerated reductions in the amount of mixed cropping and traditional land management. Indeed, changes in the expectations of young farmers will ensure that further change is likely to be characterized by increases in farm size and greater specialization with implications for landscape and wildlife. These characteristics of farm holdings are more familiar in the east regions of Ireland , UK country. As with likely to extend to the west regions as the older generation of farmers retires, although this will probably be accompanied by a trend to more farming of production needs to young farmers. So, good natural environment can provide Ireland young farmers to produce more agriculture to earn income, even who can export more rice, fruits, vegetable etc. agriculture foods to overseas. Hence, Ireland GDP will be raise if it can have good natural resource environment to provide Ireland young farmers to grow foods to sell to domestic and /or foreign agricultural market. Given the rate of economic growth, and its concentration in the east of the Ireland, UK country, it would be easy to presume that the quality of the environment is higher the further away from the mid east one goes. Thus, good natural environment is an important factor to influence the farming industry development in Ireland , UK county to satisfy their needs and to raise their quality of life nowadays.

I shall indicate New Zealand and America two developed countries to explain why which are facing environmental pollution challenge to influence their citizen's quality of life and economic growth and cause food wastage nowadays. The first country is NZ, although, New Zealand is a developed and natural environmental country, but it had been envountering air pollution annouance and noise annoyance to influence it's citizen's health-related quality of life. I shall indicate why which has this relationship between of them in New Zealand. Nowadays, New zealand population growth is an increasing demand for consumer products and urbanization have lead to concerns over the lived environments in many of the world's cities, such as Auckland, wellington cities in New Zealand. However, environmental quality is an important determinant of health, such as the bad influence of traffic-related air and noise pollution on health outcomes, specially with respect to at risk groups, both in relation to long term exposure as well as acute effect, from brief exposures. For example, cholesterol levels and in relation to myocardial infaction. Nowadays, New Zealand is encountering the high degree of air pollution and noise annoyance to influence it's citizen's quality of life. Air pollutants can be detected either visually, such as witnessing smoke emanating from a vehicles's exhaust, or by smell, such as when odorants stimulate olfactory receptors. The evidence linking air pollution to adverse impacts on human health.

Many air impacts on human health. Many air pollution health studies have focused specifically on urban area, and vehicle generated pollution in particular, as road vehicles are one of the major sources of pollution across much of the world. Elemental carbon, Nox and ultrafine particles an considered to be pollutants most strongly associated with road traffic emissions. In Auckland and Wellington cities, New Zealand , it has been estimated that 71% of summer and 21% of winter concentrations of fine particulate matter is attributable to motor vehicles. Moreover, poor town planning decisions in Auckland (and in New Zealand in general) over many decedes has meant that may people live in very close proximity to busy road and motorways within " road corridors" and so are the adverse effects of road traffic, including noise and air pollution as well as experiencing on potential for degradation in their quality of life. Such as, New Zealand is highly suitable for studies investigating the impact of roads on the health of its residents. For example, NZ, road traffic noise and aviation noist has been linked to cardiovascular disease, hypertension and ischemic heart disease. It influences NZ resident personal psychological and physical both health challenges. In fact, NZ noise increases morbidity and mortality independently of air pollution exposure, though air pollution constituted a greater burden of disease when arise exposure had a greater impac on quality of life, e.g. NZ road traffic noise and air pollution will be caused from drivers in busy time. Specially in Auckland and Wellington cities. It will influence urban and rural environmental pollution. Some retired old people who will feel annoyance when this road traffic occurs in Auckland or Wellington cities to close to their houses in transportation busy time every day.

Next developed country is America, this country's air pollution is also serious nowadays. Because traffic jam often occurs in New York, Washington, Boston etc. big cities in US. So, U.S. cities' parks and its trees have significant influence to produce fresh air to provide U.S. residents who are living in cities to breach for their body health. David J. & Gordon , M. (2016) indicated " In U.S. these urban parks are estimated to contain about 370 million trees with a structural value of approximately $300 billion." The number of park trees varies by region of the country, but which can produce significant air quality effects in and near parks, related to air temperatures, air pollution, ultraviolet indication and carbon dioxide (a dominant greenhouse gas related to global climate change). Additional open space and other vacant lands in cities, which may contain trees and other vegatation. Contribute significant additional benefits, effects of parks and open space at the city scale can vary significantly depending on the amount of parkland and amount of tree cover within the parkland.

The reasons why parks can reduce air pollution. Parks generally have lower air temperature than surrounding areas. Temperatures are usually cooler toward the center of a park than around its edges. At night, the center of a large park may be 13 degree cooler than surrounding city areas. The cooler air from parks often moves out into adjacent developed neighborhoods. This cooling of surrounding areas tends to increase with park size and percentage of the park covered by trees. So, cooler air temperature is provided by urban parks can have significant impacts on human health. During heat wave events, which can kill hundreds of people, park areas may provide city dwellers with some respite from high air temperture, particularly in the evening, during hot, sunny days tree shade can greatly increase human comfort. Because park influences on air temperature extend to developed areas outside of parks, local energy use for heating and cooling buildings is also effected.

Although, the net around effect of parks on energy costs has been by reducing temperature is difficult to estimate at least in the southern United States the effect will usually be a net annual benefit. Futhermore, large park trees will reduce winds and may provide a benefit of winter heating of buildings near the park. Although, the overall economic effect of urban trees and parks on air temperature reduction is not fully billions of dollars annually at the national scale in terms of improved environmental quality and human health.

In fact, trees and vegetation in parks can help reduce air pollution both by directly removing pollutants and by reducing air temperatures and building energy use in and near parks. There tree effects can reduce pollutant emissions and formation. However, park vegetation can increase some pollutants by either directly emitting volatile orgnic compounds that can contribute to ocone and carbon monoxide formation or indirectly by the emission of air pollutants through vegetation maintenance practices, such as operation of chain and use of transportation fuels. David J. & Gordon , M. (2016) showed "Annual pollution removal and economic benefits by U.S. urbank park trees is estimated at about 75,000 tones ($500 million) or 80 pounds per acre of tree cover ($300 per acre of tree cover). Carton storage and annual removal by urban park trees and soils in the United States is estimated at about: carton storage trees: 75 million tons ($1.6 billion), carton storage (soils) : $102 million tons of carbon removal (trees): 2.4 million tons ($50 million)". Park management is recommended by U.S. environment protection department: considering that most of the effects of trees on microclimate and air quality are beneficial for park users and nearby residents; park designs that include a variety of land cover, areas of dense trees, scattered trees and lawn are likely to provide the greatest opportunities for optimum physical comfort of visitors; increase the number of healthy trees (increase pollution removal and carbon storage); sustain existing tree cover (maintains pollution removal levels) and (carbon storage); maximize use of low volatile organic compound emitting trees reduces ozove and carbon monoxide formation; sustain large, healthy trees (large trees have greatest per tree effcts on pollution and carbon removal); using long-lived trees (reduces long term pollutant emissions from removal; reducing fossil fuel in maintaining vegetation reduces pollutant ans carbon emissions)." So, if US had many green parks, then which can reduce air pollution, also it can assist many travellers who prefer to travel to US to raise GDP travelling income growth generally.

(i) Why environmental pollution and human right abuses has close relationship to influence quality of life and economic growth as well as bring food wastage ?

In fact, environment pollution and human right abuses has close relationship. It is clear that poverty situations and human rights abuses are worsened by environmental degradation. The result can influence poor human quality of life to the developing countries' people unfairly. There are these several abvious reasons: firstly, the exhaustion of natural resources leads to unemployment and emigration to cities; secondly, this affects the enjoyment and exercise of basic human rights. Environmental conditions contribute to a large extents to the spread of infections diseases. From the 4,400 million of people who live in developing countries, almost 60% lack basis health care services, a almost a third of these people have no access to safe water supply; thirdly, degradation poses new problems, such as environmental refugees. Environmental refugees suffer from significant economic, socio-cultural and political consequences. And fourthly, environmental degradation worsens existing problems suffered by developing and developed countries. David J. Nowak & Gordon M. Melsler (2016) showed" Air pollution , for example, accounts for 2.7 million to 3.0 million of deaths annually and of these 90% are from developing countries. " Hence, our societies need to concern human right law to protect unfair treatment to developing countries people. Firstly, both disciplines have deep social root, even though human rights law is more rooted within the collective consciousness, the accelerated process of environmental degradation is generating a new " environmental consciousness". Secondly, both disciplines have become internationalized . The international community has assumed the commitment to observe the realization at human rights and respect for the environment. Thirdly, both areas of law tend to universalize their object of protection. Human rights are presented as universal and the protection of the environment appears as everyone is responsibility.

Human right and environment law can raise our quality of life because the first approach is one where environmental protection is described as a possible means of fulfulling human rights standards. Here, environmental law is conceptualized as giving a protection that would help ensure the well-being of future generations as well as the survival of those who depend immediately upon natural resources for their livelihood. So, the end is fulfulling human rights, and the route is though environmental law, the second approach places the two sphere in inverted positions, it states that the legal protection of human rights is an effective means to achieving the ends of conservation and environmental protection. Therefore, the presently existing human right is as a route to environmental protection. The focus is on the connection to influence any economy: health, food supply , housing, fresh natural air supply etc. aspects of quality of life issues. Hence, human right and environment law and human quality of life and economic growth has close relationship . We can not neglect to concern how to achieve human right law to protect our nature environment existing in our societies.

What are environmental factors affect human health in important way, both positive and negative? On positive environmental factor aspect, which can sustain health, and promoting them is preventive medicine. They include : sources of nutrition (farming, oil quality, water availability, bio diversity/bio integrity, genetically modified organisms ; hurting, fishing: wildlife, fish populations; water (drinking, cooking, cleaning,sanitation); air quality; ozone layer (protection from cancers disease etc).; space for exercise and recreation, sanitation/waste recycling and disposal.

On negative environmental factors aspect, which are threats to health, and controlling them is public environmental health. They include: environmental conditions favouring disease sectors (endemic and exotic sectors); invasive biota (visuses, bacteria etc.), their hosts and sectors; environmental disruptions: floods, droughts, storms, fires earthquakes, volcanoes; air quality: pollution landing to respiratory disease or cancers; water quality: biotic and abiotic contaminants ; integrity of water transport and intrastructure; monitoring and management of municipal, agricutural, industrial outflows to the environment (gases, liquids, solid waste), human changes of the environment that: create conditions that favour disease; disturb and release noxious levels of previously bound chemicals (e.g. mercury released becomes poison) or bioto (e.g. methane released from thawed peat contributes to climate changes, create temporary, intense, life threatening heat islands (e.g. urban heat waves exacerbated by climate change); result from nuclear; biological or chemical welfare or terrorism, disruption cased by other war and violense.

(ii) What is space and environmental technology to avoid food wastage?

For example, Cananda is a developed country and it begins to concern environmental pollution challenge to announced $3 million to support the initiative strengthening health and environment linkages: from knowledge to action. The initiative will bring together scientific, technical and socio-economic information on environment and health linkages, and transfer that knowledge to inform decision-making at the local, regional and national levels. Also, Canada is principally concerned with the health of Canadians. This involves health factors in Canada and in biologically shared health regions (shared geography or exposure through trade and travel). Supports international health initiatives, such as determining health risks throught environmental analysis of disease vectors in Africa or Asia.

How can the space and environmental factors affecting health? Environmental information and environmental management contribution to the maintenance and restoration of health. Space based environmental management factors and communications can play roles in: Environmental information is for optimising use of health resources; distribution of and access to health advice and treatment (i.e. to health staff treatment facilities; short range environmental prediction for avoidance of high risk, situations and to guide immediate health system responses. Managing acute risks, adopting to them (e.g. temporary moving of vulnerable elderly monitored; modeling of health impact of environmental parameters; prediction of long term health resource needs and environmental planning and mitigation and adaptation to global changes. Large benefits are possible from attention to environmental factors, e.g. asthma prevention, disease and epidemiology. Benefits need to be quantified. This is of particular interest and relevance to pandemics , such as malasia in underdeveloped countries, potentially saving thousands of lives.

What is space and environmental technology? It can contribute to and keep abreast of environmental health forecasts (using existing models and known parameters); prepare and deliver prospectuses for what space can do in anticipation or response; steer space programs according to real risks and real accumulative health benefits, as long technical investment, don't focus primarily on threats that may have high emotional impact , but are of low actual risk; position space technology and the canadian space program in people's winds, aggressively and realistically, as a first line contributor to foresight and preduction, long term maintenance of well-being and prevention of factors of ill-health ; ongoing delivery of health services and management of current health factors and potentially capable and ready to respond in health emergencies. Finally, making the full business case for investment in space technology and space program contributions relative to the full and public and private cost of health programs. This connects not only to GDP raising, but to indicators of quality of life to any countries.

Reference

Cornelia, B.F. (1999) Rural development news, the North Central Regional Center For Rural Development vol. no 24 , IOWA.

David J. Nowak & Gordon M. Melsler (2016) " Air quality effects of urban trees and parks." National recreation and park association, USA.

De Hollander, A. E. M., J.M. Melse, Elebret & P. G.N. Kramers (1999), " An Aggregate public health indicator to represent the impact of multiple environmental exposures" Epidemiology: 606-617.

Felce, D. and Perry, J. (1995). Quality of life: A contribution to its definition and measurement, vol. 16, no.1 pp: 51-74.

Los Angeles Country Department Of public Health (2016), Country Health Ranking Model, Retrieved From www.countryhealthrankgings.org/our-approach. USA.

Melse, J.M. & A.E. M. De Hollander (2001). " Human Health And The Environment", background document for the OECD Environmental Outlook, OECD, Paris.

McGregor, S.L. T., & Goldsmith, E.B. (1998). Expanding our understanding of quality of life, standard of living and well-being. Journal of family and consumer science, 90(2), 2-6, 22.

McMichael, A.J. M. Mckee, J. Shkolnikov and T. Valkanen (2004), " Morality trends and setbacks, global convergence or divergence?", Lancet 363, 1155-1159.

Yale Center For Environmental Law And Policy (2006). Environmental Performance Index. Data available on-line at http://epi.yale.edu

The failure of human education method
influences consumers to change food and energy
waste behaviours

Can artificial intelligence analyze human education method is the main factor to influence consumers to change food and energy waste behaviors ? I shall indicate the reasons to explain why human education method can not persuade to change consumers and manufacturers to do food and energy waste or loss behaviour easily.

The first reason is that , food waste is a serious ethical, environmental and economic problem of excessive consumerism. Usually, food waste occurs in all phases of the food supply chain, starting with producers and ending with consumers. There are several factors which contribute to excessive qualities of wasted foods. Some are related to current quantities of wasted foods . Some are related to current production systems and product commercialization, including food quality and security norms, others are more personal like people's food habits, awareness, values and consumer attitude in regards to consumption and food waste.

So, changing consumers and manufacturers whose wrong food consumption and excessive quantities of wasted food behaviours. It is difficult to avoid from only education method , because human's eating habits or food habits are very difficult to change to reduce to do food waste behaviours in their eating processes easily. Education is only teaching how to avoid food waste knowledge promotion or persuasive method. It can not provide useful food waste methods to assist food manufacturers how to present to avoid excessive food quantities waste in food manufacturing processes , e.g. efficient operative production as well as it can not assist consumers to understand how and why the reasons for food waste, why to do the right food consumption behaviours, eating attitudes, the most reason level of consumer knowledge and opinions with respect to food waste as well as perception about the quantities of wasted food in the consumer's household. Otherwise, future (AI) technology can assist food manufacturers how to avoid and to cause food losses in food manufacturing process as well as it can assist food consumers understand they ought how to do to change wrong food habits to right food habits in order to do the right daily eating behaviours daily more influentially.

The second reason is that, in order to reduce consumer individual food waste in developed countries, there is a need to understand the factors which shape consumer behaviours. Future (AI) technology can be used to research and analysis of customer behaviours, knowledge and attitudes of people including analysis of different developed countries' consumer behaviours for food. The objective of (AI) technology research is to determine food waste attributes, daily food routines, shopping routines, planning as a predictor of food waste and policies varied in terms of household characteristics in each country.

The (AI) data gathering results can give qualitative information about food waste including data on frequency of wasting food and reasons for wasting it, which can be based on each country's consumers' food habits or eating behaviours analysis. The (AI) gathering data concerns consumers play a crucial role to combat food waste via their own households. It is important to find solutions that are relevant to food habits more easily.

Hence, (AI) technology can gather data to understand food waste problem to household level by picturing main causes affecting food waste giving deeper insight into consume behaviour throughout the every time of right food purchasing number, storage, preparation, consumption toll disposals. So, future (AI) technology (big data gathering tool) can analyse origins and quantities of food waste quantifying the scale of the problem, including all causes and food waste influence as well as to understand the variety of factors which can cause influence food waste behaviours more accurately.

The third reason is that, future (AI) technology can also be applied to solve food loss challenge. Food loss is considered to be food that gets spilled before it reaches its final product or retail chains, which occurs at production, post-harvest, and at the food manufacturing or transportation processing stages. It is also caused by poor infrastructure , and logistics, lack of technology , inefficient skills, knowledge and management capacity , it may be accidental or intentional , ultimately leads to less food available. So, it seems that education method is difficult to avoid food losses for food manufacturers. However, future (AI) technology can be applied to assist farming agriculture to grow crops to avoid food loss easily in good climate or food growing environment.

Instead of (AI) technology can be applied to avoid humans do food waste or food loss behaviours easily. It can also be applied to avoid energy waste aspect. I shall explain why (AI) technology can change or influence householders and manufacturers' energy using or consumption behaviours to save more electricity, gas energy at homes, offices or plants or shopping centres etc. working or entertainment places more easily or effectively to compare energy waste educational or learning method. I shall indicate the reasons as below:

(AI) technology is the more quality training data to a (AI) data gathering energy sector, which has access to the better , the (AI) can be integrated into its everyday operations. The energy sector is one of those industries with a wealth of data that is for machine learning and its comparison technologies to assist humans to avoid to do energy waste behaviours.

The commercial opportunities from using (AI) in this energy industry come from the technology's real-time optimisation, predictive analysis and forecasting power. For example, a number of New Zealand energy distribution companies are working with postgraduate students in big data and machine learning to analyse information gathered from both their networks and smart meters. Between 75 and 85 per cent of data in this sector is structured and machine learning can be used to analyse it, platforms using unsupervised learning techniques are well-suited for energy detections which can boost the efficiency of utility operations to reduce optimising energy cost. Machine learning algorithms are being used to understand vast amounts of data to predict mining, drilling and power generation failures and then to recommend tailored maintenance based on the potential issues.

SO, (AI) systems can monitor the emission of nitrogen oxides from gas turbines and vary the distribution of fuel to maintain the required levels of energy generation. Scaled across the network, this is a genuine opportunity to better control energy useful cost to avoid energy waste. Moreover, machine learning can also manage energy to use within complex systems. For example, Goggle's deep mind (AI) achieved a 40 per cent reduction in energy used to cool the company's data centres, even after human engineers had supported optimised the facility's energy use. ON a bigger scale, machine learning can be employed to manage the use of resources, such as water energy in " smart cities" to avoid to cause water natural energy power is excessive waste.

On benefits to energy end-users, based on real-time usage data, machine learning algorithms and supervised learning can improve energy management , even in power plants. This in turn optimises total pricing for energy consumers and provides opportunities to offer promotions based on energy customer demographics. SO, machine learning is used in some of these products as part of a deeper effort to provide customers with more choices about how to produce, consume and store energy and encourage them to do in many energy saving right behaviours.

(AI) has the ability to transform resources distribution and compensate for drastic fluctuations in energy demand. The technology will also go beyond just pricing and distribution to respond directly to the dynamic needs for energy, water power and natural drinking water and waste management. Through an ecosystem of " smart city", water and electricity can be distributed from many small energy producers and be personalised to specific regions and time frames. Each energy producers could use techniques which are particularly suited to time series data . This network would take in and learn energy user behaviours and use the information to manage the energy or water power or drinking water supply . A energy producer can then sell excess energy capacity back to the grid, maximising efficiency and reducing energy wastage to energy consumers.

SO, in the future, the key (AI) technologies in energy sector, they will bring high impact on machine learning, natural language programming, chat bots which will be applied to machine learning technology in energy saving sector to avoid energy wastage when energy consumers use energy daily. With the high volume of data available, global energy businesses can readily continue develop their (AI) technologies. Better resource allocation, improved customer satisfaction are obvious benefits as well as the most important influence is that (AI) energy saving consumption or using technology can impact energy end –users to avoid to waste too much excessive energy when they are using energy at homes or any public places or private factories or offices manufacturing places for every day necessities. So, it can explain why (AI) technology can impact energy end-users energy saving and using behaviours to be improved.

Next, two chapter, I shall explain how (AI) technology big data gathering tool how it can help energy end-users and food consumers how to impact to avoid to do wastage behaviours more influentially every day.

How (AI) technology impacts food consumers
and food manufacturers food eating habits
or attitudes to avoid wastage

Can artificial intelligence impact food consumers' eating habits to be influenced to change ? Future (AI) technology (big data gathering tool) can be used to gather data concerns to supervise or manage or control these below food consumers and food manufacturers' food waste or food loss behaviours in order to find the main reasons to cause their food waste or loss wastage behaviours in order to achieve to prevent or avoid their wastage behaviours occurrence again more easily. I shall explain to these aspects as below, they include :

On (AI) auto-supervised food consumer individual food waste behavioural aspect:

Firstly, for the original food wastage supervised, the original food which included food in unopened packages behaviours, which was thrown away because it passed the expiration date including products to like cheese, yogurts and other daily products, loose fruits and vegetables which became rotten and was never used. (AI) technology can follow these waste food gather number from different countries' supermarkets, food stores to gather waste food number in order to carry statistic analysis to find the reasons why these kinds of original food, cheese, yogurts, fruit and vegetable wastage number has increase to every month in order to attempt to find the reasons to cause wastage food behaviours , it is either caused by either food manufacturers' food losses (good manufacturing process negligent factor) cause or food consumers' food eating habits cause in order to find the solution methods to change their wrong food loss or waste influential behaviours to the food manufacturers as well as the wrong food habits or attitudes to the food household consumers more easily and accurately.

Secondly, for another kind of wastage food is partly used food supervised, the food which could have been opened or started, but was never finished. (AI) data gathering tool can attempt to follow different countries' food rubbish to gather the number data to find how much rubbish number is belonged to the country's wastage food is partly used food or the food which could have been opened or started, but was never finished to find the main reasons (factors) why they caused those kind of food wastage number is increased from food consumers; eating habits or what the reasons (factors) caused their number is decreased the country's food consumers' eating habits in order to find the most effective or accurate methods to solve this kind of food wastage behaviours from the country's food consumers.

Thirdly, for the another kind of food wastage is leftover supervised, which consist of food left or the plates or were cooked in big amounts which ended is not being eaten. (AI) technology big data gathering tool can gather global these kind of food waste number concerns every householder's food left on plates or were cooked in big amounts to every country, but not being eaten number from their rubbish.

TO attempt to find what factors(reasons) influence their food habits to do food left on the plates or were cooked in big amounts , which are nor being eaten food habits. SO, it can follow this kind of global food wastage increasing or decreasing number every month statist data to attempt find what factors influence this kind of food wastage to household food consumers to be either decreased or what factors influence this kind of food wastage number to be increased to conclude the more accurate food behavioural wastage judgement for this kind of food wastage habits to every country household food consumers, e.g. life habitual factor, food price factor, food perishable factor, climate influence factor etc. different kinds of external factors to cause every country's household food wastage behaviours.

Finally, the final kind of food wastage is that preparation residues, (vegetable peels, egg shells) supervised, this kind of food wastage could potentially be still used and not known away by global every householder food consumers, they can not be avoided to waste before cooling, due to the householder feels these foods are not fresh to eat. So, the householder chooses not to cook it to eat.

However, (AI) technology can gather data concerns different countries' householder fresh food purchasing habits ,e.g. per week or per twice week or per day fresh food purchasing frequently and fresh food purchasing number, such as vegetable peels etc. fresh food wastage rubbish number in order to find what the main factors (reasons) is (are) to cause the next month fresh food wastage number to be increased or decreased to every country householders in order to conclude the more accurate or reasonable wrong fresh food wastage habitual behaviours to cause different countries' fresh food householders' fresh food wastage behaviours habitually.

In conclusion, basing on above evidences, I can explain why future (AI) technological big data gathering tool can be applied to find solutions to avoid that food consumers to do food wastage behaviours habitually. It also explain why food wastage education method is only knowledge concept to educate to let public to know. In fact, food consumers can choose either do or not do to avoid food wastage in their eating habits daily. Otherwise, (AI) big data fathering tool can be applied to attempt to gather global householders (food consumers) their daily eating habitual data in order to achieve the more accurate and predictive householders (

food consumers) their eating habits or eating behaviours analysis and concludes the most efficient and effective solutions to avoid global food wastage number to be raised to every country household food consumers. So, (AI) technology can impact global householders (food consumers) eating or food habits to be improved more better to compare education method.

On (AI) auto-supervised food manufacturer individual food loss behavioural aspect:

How (AI) technology changes food manufacturers cause food loss in their food manufacturing processes, I shall indicate as below:

How to apply (AI) technology to avoid vegetable, fresh fruit loss to global farmers' fresh vegetable , fruit excessive wastage loss increasing number when their crops growth process in global agricultural sector? (AI) will enable significant and valuable new solutions to avoid fresh crops, fruit, rice, vegetable food loss in their growing or irrigation process.

The internet of things (OIT) will assist (IA) in future intelligent agricultural systems, fuelled by large volumes of data acquired from images, videos and IOT sensors. For example, it be applied to water and sprays in agricultural sector. (AI) is smoothing the way for new levels of optimisation on our farms and across all horticultural activities. The automated irrigation systems are getting and the transporting of water to specific places. It is based on the real-time needs of plants. Moreover, (AI) techniques using IOT and sensors to analyse what's happening across many hectares of farming land in real time will enable improvements in predictive modelling.

Farmers will be able to check the advantages of specific phenotypes , or traits in certain growing over time . Predictive modelling will also help them forecast pest resurgences, dramatically preventing yield losses and reducing farmers' dependence on chemical pesticides to let vegetable, fruit, crop can be grown healthy.

For the university of Waikoto , NZ example, researchers are applying machine learning to near infra-red images of soil, meaning the soil does not have to be sent to the lab. This will enable farmers to apply fertilisers much more efficiently. Moreover, robots will be developed to roam between strains of needs. These bots are high enough that do not damage the soil, and because they release herbicides only onto the weeds. They are also doing more environmental protection behaviours. Many uses of (AI) in agriculture are focused on reducing the biological and ecological damage caused by inefficient use of pesticides.

For meet waste loss reducing aspect, (AI) technology can be applied to animal health monitoring . It can also be used to improve efficiencies in livestock management by optimising feeding and dispensing of medication. The (AI) technology can constantly monitor livestock , e.g. pigs, cows sheep animals movements, eating patterns and health and immediately flag animals that are showing unusual behaviour or reduced welling. They can then be treated quickly before they spread infection. So, farmers benefit is from cost reduction through more targeted use of antibiotics when also improving the treatment of livestock.

In the simplest terms, images are constantly captured and pre-processed through detection for frequency and density to livestock eating animals' eating behaviours and living and health conditions in order to provide health pork, beef, sheep meet to consumers to eat.

However, (AI) can be applied to crop seed or fruit improved growing or better irrigation aspect, for New Zealand kiwifruit agricultural irrigation case, yield well ahead of scheduled harvests. New Zealand kiwifruit growers have had to manually count fruit over certain areas and then to achieve the more accurate kiwifruit supplying number of consumers' demanding number scheduled harvests to satisfy New Zealand itself country's kiwifruit consumers' needs , even overseas kiwifruit consumers' needs. The agricultural sector would not know until the kiwifruit product hit supermarket shelves whether its spot sampling had been correct. Any miscalculation could cause kiwifruit waste loss. For example, if NZ kiwifruit farmers predict China kiwifruit consumer number will be on million kiwifruit consumer number in this year. However, they miscalculate the wrong kiwifruit supplying number either their needs are lesser. Then, it will occur shortage kiwifruit number to export to Chinese kiwifruit consumers in this year or their supplying export number are more, then it will occur excess kiwifruit number to Chinese kiwifruit consumers. So, (AI) technology can help NZ kiwifruit farmers to predict every country's kiwifruit needs in order to supply the enough kiwifruit number to every countries' supermarkets or fruit stores to sell. SO, it won't cause NZ kiwifruit excessive or shortage challenges occur more easily.

Hence, (AI) enabled technological tools can observant or supervise every countries' fruit consumers' eating habits to predict whether how many fruit number that they will need to eat every year more accurate. It is as simple as a smartphone to video along on trailers can help provide better estimates. The device videos , the orchard on –the-go and can then produce global fruit consumers' fruit eating habits to predict their fruit consumption behaviours more accurately in every year. From these video-based machine learning systems can detect more better and therefore count seeds and fruit months in advance. So, having these (AI) predictive crop or seeds growing number technology to predict whether is enough insights in advance would also enable harvesters to undertake section-based optimisation, improve food safety and direct fertiliser to specific locations. Global faming suppliers can better manage their crop , fruit seeds, vegetable seeds growing process and time to control or predict when they can be grown to sell when it can reaches the mature stage , even meat pricing and revenue forecasting with supermarkets or food or fruit store retailers accurately book shipping logistics and storage and reducing waste.

So, key (AI) technologies will extreme impact agriculture to change farmers' vegetable, fruit, tomato, potato, crop, and livestock animals feeding methods or behaviours to be improved better by machine learning, drones, computer vision. IOT robotics, satellite , data influence.

How to putting (AI) work in the vineyard for grape fruit growing better? For Lincoln Agri. Tech. a research and development company owned by Lincoln university , NZ case example, it is developing an (AI) solution which can make early season predictions of vineyard harvests. (AI) technology can help NZ grape growers and wineries to predict their grape yield each year. SO, (AI) technology can help them to do grape yield prediction work. A large number of manual workers do sample grape bunches work. SO, (AI) technology is working on creating a system that instead uses electronic sensors to accurately count grapes for NZ grape fruit farmers. The sensors will capture and analyse grape bunches within individual rows, and access the number, sizes and distribution, feeding these different kinds of grape number data into computer algorithms in order to predict grape yield at harvest time to calculate the different kinds of grape yield to different countries grape consumers' needs more accurately. New date will also be added to the (AI) computer system each year, leading to continuous improvements in the model's accuracy as more information is gathered under different conditions. Hence, (AI) system will enable NZ grape growers to accurately access differences in yield , not only between regions or vineyards, but also blocks and rows. Over the long term, site-specific grape yield prediction will help reduce costs by enabling better planning both in the vineyard and in NZ grape market, even overseas grape market both. This (AI) technology will benefit the agricultural industry by supporting better crop or seed management, smoother processing and fruit , vegetable, crop market based on capacity to supply the more accurate number.

Global farmers can apply (AI) technology to improve agricultural water system to let crop, fruit, vegetable to grow more easily to avoid waste loss number rises. Applying on farming sensors and other data to provide farmers with daily recommendations around nitrogen application and water and effluent irrigation . It can be accessed via smartphone, take the guess-work out of interpreting the large amounts of data farmers need to consider before making an irrigation decision.

(AI) agricultural water irrigation computer system can expand into nitrogen application management and water irrigation scheduling. Optimising the amount and timing of effluent are key components of ensuring farm sustainability and resource consent compliance, minimising leaching and optimising pasture growth.

Future (AI) technology just-in-time water management can provide enough water supply to satisfy agricultural fruit, vegetable, crop etc. food irrigation need to global different farming lands more easily. In farming industry, enough fresh water irrigation need is very important to influence any fruit , vegetable, crop etc. food growing process successfully. (AI) technology can be applied to this farming land water supply aspect. It can measure when the farming land needs how much water supply to provide to the farming land's fruit, vegetable, crop growth more easily. Either when the farming land does not need water supply, or when the farming land has still have enough water in the farming underground land. For example, a Florida water company is using artificial intelligence to reduce withdrawal of water from the area aquifer in Jacksonville, USA. Tis (AI) water supply measurement system can forecast water consumption , then monitors, regulates and adjusts supply in real-time , providing a just-inOtime water supply to the farming underground lane. This minimises wall production during peak hours, optimises reservoir storage, and reduces the number of pump starts required, lowering energy consumption and maintenance costs.

The (AI) water supply management system can meet the fruit, vegetable crop growing demands of these food consumers needs and reduce the need to big new wells and preserve for un-predictive water supplying need for any fruit, vegetable, crop water need on any farming lands. So, every farming land will have accurate water supplying , it won't have excessive or shortage water supply challenge to every farming lands, if the country's farmers chose to use this (AI) water supply management system.

In the future, (AI) technology advantages to be applied to food waste aspect, it can influence: assisted farming to provide enough water supply to irrigate the most accurate and measured water to every farming underground lands to let any fruit, vegetable, crops' seeds to have enough water supply to grown rapidly, reducing open-sea fishing, considered use of farming land energy supply or affordable and clean energy supply.

In the future, artificial intelligence has the potential to let any fruit vegetable, crop seeds grow up successfully. It can be used to analyse seed genetic data to create crops that can thrive and adapt in any sudden changing environment or weather in order to keep the fruit, vegetable, crop seeds can still have large adapting effort to grow up in any bad conditions. It can increase effectiveness of food supply chains through the use of (AI) technologies to drive insights that improve any fruit, vegetable, crop growing up efficiency and reduce waste in any farm. It can also give recommendations to support for farmers to choose the best decision making in order to increase in productivity. Instead of these , it can reduce the food loss in growing or manufacturing process. It can also help human to choose the healthcare diets. It includes to mine health care records to improve quality of treatments and provide better and faster health diets. It can analyse of large scale genetic data sets to help create new types of treatment and precision medicines customised for individuals, providing expert assistance in diagnosis especially in repetitive tasks, such as analysis of images and large bodies of research information, providing immediate first line consultation to improve waiting times to see a doctor, reducing pressure on frontline staff by using robots in healthcare, speeding up the development of new drugs allowing treatments to reach whose who need them more quickly, improving learning outcomes through the analysis of data about individual learning, social and learning contexts, and personal interests, providing virtual monitors for learning by integrating modelling, social simulation and knowledge representation, providing lifelong learning companions that help the learner to adapt and build new skills throughout their lifetime. Hence, it can bring the actual life knowledge to let food manufacturers, farmers and food consumers to learn how to avoid to do food wastage or food loss behaviours in our daily life experiences. Moreover, it can help farmers , food manufacturers to learn how to optimize energy generation and reduce environmental impact through analysis of operational and environment data as well as helping food consumers or food manufacturers to find the most affordable and clean energy in efficient ways od deals when they are cooking or manufacturing foods by using (AI) agents. Even, it can improve actions to fight climate change through better modelling and analysis of large or complex data sets in order to generate better insight to help to improve the sustainable management of land resources, e.g. soils, forest, biodiversity and allow greater understanding of the impact of better farming land use choices for fruit, vegetable, crops ' seeds growth through global agricultural industry development predictive analytic and machine learning.

In conclusion, based on above different (AI) technology predictive function s, so it explains why it can replace education method to avoid or reduce future good or energy shortage challenge more easily. It is due to food consumers or food manufacturers' food waste or loss behaviours which can not be controlled to avoided to cause easily in their food manufacturing or crop growing processes as well as food consumption processes. (AI) green data revolution will create a smarter, more flexible food waste controlling system as more data is created and shared between food supply chain partners and food consumers. It can bring positive benefits to agricultural industry, such as food factory manufacturing automation, intelligent food packaging, food waste or loss risk analytics in manufacturing processes , food supply chain number forecasting, food product personalisation and new avoiding food waste or loss ways of engaging with food consumers.

Finally, I shall discuss how (AI) technology can gather data to bring food waste or food loss threat consequent message to let humans (food consumers or food manufacturers) to understand how to change their eating waste habits or food manufacturing loss behaviours more easily as below:

Future (AI) technology can gather global food manufacturing and food consumption data to conclude one accurate food system trend model to let us to influence our negative or wrong eating habits or food manufacturing behaviours to improve to more positive in order to avoid food loss or food wastage causes easily.. The food eating or manufacturing behavioural system trends model can be as below:

It includes how to consider a number of trends that will influence the food system over the coming decade, focusing on small number of " key trends" in agricultural industry and supermarket, food stores, food grocery and food manufacturing industry, fishing industry, restaurant and hotel food supply industry, these trends which focus on a number of issues across the natural environmental, social and economic influences , due to future food waste or food loss consequent causes.

Changing social norms and working practices continue to influence the frequencey and format of food consumption. Predicting how the soical media influences to the way householders make food purchasing and dining decisions to influence food consumers' food waste behaviors from online. Predicting how the widening wealth gap and aging population are altering household food shopping and eating behaviours changes. Predicting explosion of data enabled technology concerns how the amount of data is growing at an predictive rate. The offers potential for a smarter, more responsive reducing or avoiding food waste behavioral system to food consumers or food manufacturers. Predictinv how the positive changes or influences of tackling major public health and environmental challenges are caused from humans (food consumers and food manufacturers) whose food waste or food loss behaviours. Predicting food system (big data gathering method) concerns food consumers or food manufacturers' frequency of food waste or food loss behavioural data to adatp to an uncertain operating environment as well as to find the methods how to fight the risks of climate related shocks to the food shortge challenge.

Hence, the (AI) data gathering aims to achieve these missions as below:

Learning how to grow in most technology is driven a revolution in how the entire food system operates from a better understanding of land resources to automated factories and kitchens. To bring data enabled technology that will be become cheaper and more accessible all the time, but the avoiding food waste system will fully capitalise on the benefits over tehe next ten years. This avoiding food waste system will be explored to companies, households and food waste policymakers seek to make better use of data. They include as below:

(1) Increasing number of devices connected to the internet and number of social media users to learn how to avoid to do food waste or food loss behaviours daily.

(2) Decreasing the cost of data storage technology and the size od data enabled devices.

(3) The food waste avoiding system can deal with th increasing complexity and sudden changing environmental , social and economic systems to avoid the failure to respond proactively to these new challenges to bring energy, water and food shortages consequences.

(4) Increaing efforts to fight the impacts of extreme weather events, agricultural pest ranges to the food avoiding waste system interconnectedness.

(5) Understanding how to increase soil health, crop diversity, global nutrition and quality and availability of water resources.

(6) Understanding how to fight climate change, it brings significantly affect the fruit, vegetable, crop's seed growth to the improved better in agricultural industry through its impact agricultural yields, agricultural food changing prices influence, reliability of supply, food quality and food safety for ensuring long term food security and supply chain (food transportation process).

(7) Developments of the use of advanced monitoring agricultural systems to increase input efficiencies and anticipate food production risks, such as adverse weather.

(8) Skills for future food wastage or shortage challenges, learning how to fight climate change challenges.

In conclusion, (AI) preventive food wastage or shortage technological system will need to gather the data concerns when global climate and environement sudden change and finding anywhere the global most suitable farming lands are located in order to provide the right agricultural farming lands to let fruit, vegetable, crops' seed which can grow rapidly, finding anywhere the lands are located which have enough natural water supply in order to help us to solve food shortage challenge in order to irrigate enough water to farming lands to let agricultural foods grow more rapidly, and finding anywhere have enough weeds or food for pigs, cows, sheeps livestocks on the land to eat. So, above these are future (AI) preventive food wastage or shortage technological system will be invented to solve any one of these agricultural seeds growth and livestocks feeding challenges.

VI

AI seventh stage development

(AI) Technological Space station market development

In the future, space intelligent control systems can be applied to space in these aspects:

Firstly, in space communication aspect, when spacecraft have a special need for (AI) system, due to communication delays and other factors, much of (AI) control system development is conducted for earth-based applications. As a result, (AI) communication system can let spacemen communicate to earth space station staffs between of them more easy and in short time communication in possible.

Additionally, ground support systems are required to facilitate communication with and command of (AI) controlled and other spacecraft. Future, numerous successful missions or which are controlled by (AI) technology have been launched. Thus, future many space missions seem to be controlled by (AI) control systems are better than none or lacked (AI) control systems assistance.

Secondly, in space exploration aspect, in the future, it is possible that fully autonomous spacecraft control has been confirmed in a limited capacity and appears to build promise for reducing any space exploration mission cost, increasing scientific returns and allowing the operation of more complex multi-vehicle missions from possible future of artificial intelligence control systems in space. Thus, future it is possible that (AI) control systems can been automated to assist any space scientists to solve problem during their each space exploration process.

Thirdly, in automated ground based testing system aspect, in the future autonomous spaceflight will start on the ground. For each spacecraft that is sent into space to explore, numerous concepts must be proposed, developed and tested on earth to ensure a successful mission.

Can future (AI) control systems be developed to be applied to automated test any spacecraft flight systems to ensure which are safe to leave earth to fly to space? Is (AI) automated system testing tool more accurate to achieve the testing safe measurement result to reduce the space flight risk than human testing effort? To answer these questions, scientists need to attempt to compare (AI) testing system and human effort to test many times of spacecraft flights in order to confirm whether (AI) testing system will be more accurate than human effort testing to every spacecraft equipment before it will fly to space.

Fourthly, in autonomous space exploration problem solution aspect, when a spacecraft encountered a somewhat unique problem as it moves further and further from earth. The increasing distance makes it take from ground-based controllers. As such, a spacecraft that will be any significant distance from earth most have some problems need to be able take collision avoidance actions and to reestablish communications with earth should they are lost.

The use of (AI) autonomous technologies started with meeting these basic requirements and performing actions (such as docking), which required too much. However, (AI) technology will have possible to be implemented because it can make missions better and less expensive. When, (AI) technology is participated to solve any unpredictable challenges when any a spacecraft leaves our earth to fly to space for future mission plans. It is likely the future existing space site networks will expand and some of which may be in remote locations. As such, there is an even greater need for (AI) automated challenge solution technology to facilities the operation of these various space stations within budget.

Consequently, future (AI) technology can be used in space on communication, space exploration, automated ground based testing, autonomous space exploration problem solution aspects. Nowadays, human is waiting when (AI) technology can be invented to assist space development, such as the day comes.

1.1 Challenges to install (AI) automated system in space stations

I shall indicate one challenge case how Earth space station installs (AI) automated control system to assist future any each spacecraft to achieve lot missions more productive and efficient. Ton install (AI) automated control system, which will follow two challenge views to achieve in success as below:

The first challenge view aims to apply (AI) automated system to increase the quality and productivity of each space mission current experience. However, proponents of this challenge viewpoint believed that the (AI) expert system should be developed to allow and experienced operator to make better flight decisions. They do not believe future (AI) artificial intelligent automated control system can make more accurate decisions or operator to reduce or avoid sudden accident occurrence chances during their every space mission trip.

However, (AI) automated control system scientists believed that this could be done by developing a (AI) automated control system capable of continuously analyzing all incoming telemetry to a depth that would be impractical ever for an experienced operator. Moreover, the (AI) automated control system scientists felt that the potential dollar value of this (AI) space automated control system investment is difficult to quantify. They indicated that a single good decision which allowed completion of a several hundred million dollar mission would clearly pay for a large amount of (AI0 automated control system work, but it is difficult to say that the

good decision was completely the result of the expert system.
Developing an (AI) space station expert system needs to meet this goal, such as : Reducing every time space mission flight accident occurrence as well as assisting human space flight operators to control spacecraft equipment to raise every space mission trip efficiency and productivity. So, it requires the incorporation of deep knowledge about a given spacecraft monitoring task.
Hence, the first challenge to (AI) expert system to be how to install to space stations, it is the unpredictable (AI) expert system investment amount spending and lacked depth knowledge skill of personnel each (AI) expert operator. It is only one (AI) expert system research stage to future space stations. It still needs time to research how to invent (AI) automated control system to assist space stations to implement productivity and efficiency as well as deducing space accident occurrence chances to future every space mission.
The second challenge view was how to use (AI) expert systems to allow each less experienced space flight personnel operator to perform at the level of more senior personnel. That has a measurable cost benefit that it allows shorter training times. This viewpoint requires an emphasis on breadth of knowledge as well as improved human-computer interfaces. Thus, the second challenge is unsure achievement of shorter training times measurement to assist less experienced personnel space flight operators from (AI) space expert systems assistance to achieve shorter training time productivity and efficiency aim.
Consequently, (AI) experts need to solve these two challenges to avoid space station operators feel worries whether (AI) technology can really help them to work in order to achieve (AI) automated operation control system achievement to space stations installation implementation for every space mission plan in success.

- How can artificial intelligence be applied to adapt the nature of space and satellite industry

The nature of the space and satellite industry how to require machine intelligence and assistance to launch, operate, maintain, control, repair and ensure achievement of any space missions.

Some examples of potential (AI) application include as below:
(1) Remote sensing and monitoring to predict how and when space environment changes, spacecraft security, and space mission location or destination tracking.

(2) Communications between ground and space as well as from satellite-to-satellite (in the case of multi-satellite stations), using radio frequencies, optical -laser communications, radar along with growing complexity of satellite-to-satellite handoffs between satellites in different orbits.

(3) Robotics in space transportation, include mission extension vehicles, space docking, satellite health monitoring manned space vehicles support(including health safety, medical , analytics, repair) and spacecraft to make their own decisions to explore, learn, identify, adapt during missions and carry out repairs.

(4) Data analytics include gathering large amounts of information and how that information can be used from national security , data privacy to assist space mission organizations to carry on any future space research analyses.

(5) Reusable launch and manned vehicles include sophisticated (AI) for return to Earth for completion of mission, e.g. researching asteroid mining resource includes analytics of substances discerned from asteroid examples, and remote mining of the same.

(6) Remote missions to Mars or Moon planets and beyond (and a board variety of information transit, and return). Satellites as alternative to terrestrial-based systems include cloud computing, cross-border broadband services, and other multi-data and information transfer.

- What is (AI) unmanned aircraft system?

Unmanned aircraft system is a kind of commercial transportation tool to provide benefits in terms of safety and efficiency to carry on implementing future every flight space mission aim. Advances in (AI) and machine learning technology are allowing it to see and act like human pilots, and to process huge amounts of data in real time search and rescue missions, e.g.

allowing farmers to be more efficient and environmentally friendly inspecting power lives and cell tower, survey and mapping of land or performing package delivery searching job duties. (AI) is allowing more automated , safer and efficient in Earth land, e.g. farming land and space planets' land, e.g. Mars, Moon planets.
The applications of (AI) in Earth or space land data gathering industry are limited only be data analytics for space or Earth land industrial inspections to navigating, future space warehouses more efficiently to assist spacecraft operators to transport any things between space warehouses easily in any space planets ' lands.

- Real-time data analytics

(AI) can collect and process huge volumes of data in real time. For example, gathering space environment or space planet land imagery that is used to take humans' hours, days or weeks to review and analyze is being automated by (AI) that strategically determines that kind of data and images are important enough to collect, and can simulate a human looking a lot of any space images in short time clearly. For example, when one spacecraft needs to perform inspections whether its inside machines or equipment which parts to be needed to repair, due to they are sudden damaged in any non-predictive accident event during the space mission journey. Then , it can use a variety of onboard sensors (AI cameras) to inspect track conditions and identify defects that are invisible in the spacecraft. Once detected, (AI) robots can be used to provide recommendations on what, if any maintenance may be necessary.

- (AI) robotic detective system

(AI) robotic detective system can assist spacecraft to avoid obstacles, such as space stones sudden collision in non-predictive space environment, when space stones are flying toward to it suddenly. So, (AI) robotic detective system can be capable of any flying autonomously space stones with human intention, and this will require (AI) robotic detective system to help the spacecraft to have able to sense when the space stones obstacles will fly toward to the spacecraft direction and react in time to avoid a collision. Hence, computer plus (AI) machine learning is helping any spacecraft navigate more effectively in dangerous space environment. (AI) robots are enabling the spacecraft will fly safely, even in dark, space stones obstacle filled space environment or beyond the obstacle reaches of GPS or other methods of Earth stations communication connectivity in dangerous space environment,

- The ethic challenges to (AI) space robotic learning system in space environment application

Firstly, space environment situational awareness challenge, (AI) needs have enable better space environment situational awareness and changing the ways are able to interact with things in any dangerous space environment. In the not-too distant future, (AI) technology will enable fully autonomous operations. So, (AI) space robots have ultimate safe responsibility to every spacecraft operators on aircrafts and on the human spacecraft pilots, when they depend on (AI) space robots to assist them to work in dangerous space environment.

(AI) space robots have responsibility to assist spacecraft operators and spacecraft pilots to make accurate judgement. It will raise important policy questions regarding the removal of human judgement from the (AI) space robot's judgement. Human spacecraft pilots make not only safety-related decisions, but in certain circumstances, especially emergencies, moral and ethical decisions, such as whether to crash space rocks sudden fly toward to the spacecraft. So, (AI) space robots will be needed to learn from space flight experience and use that learned knowledge to make appropriate moral and ethical judgements in dangerous space environment.

Hence, it brings this question: Do spacecraft pilots or/and spacecraft operators or (AI) space robots have legal responsibilities to maintain the security of any space mission flight as well as to ensure that its automation, space navigation, and space communication systems are good to be used?

So, (AI) space robots bring with its significant new issues involving (AI) spacecraft automated control system product liability, data privacy, intellectual property. The bound liability issues for risk and/or space mission business model includes allocation of responsibilities and costs for compliance between the spacecraft pilots/spacecraft operators and (AI) automated control system itself.

The standard allocation of known or anticipated risks between parties will separate provisions for allocating unknown rocks through on every space mission flight. Thinking these issues , through it is critical, and it may b e to (AI) space robots and spacecraft operators/ space pilots advantage to set cost and liabilities expectations, rather than to later dispute resolution. All reasonable scenarios should be approved when every space mission trip is planned to implement.

- Artificial intelligent robotic space mission or space tourism strategy

When one space exploration or space tourism organization can have one good strategic plan, it will have more accurate effort to predict either space tourism consumer individual leisure behavior or space exploration mission more successful. However to achieve artificial intelligence technology strategy to any space exploration mission or space tourism?
I believe that artificial intelligence technology strategy to natural language, transportation, client service, education etc. (AI) businesses implementation. It must be different to artificial intelligence space exploration mission or space tourism businesses implementation. However, artificial intelligence space exploration mission and space tourism both strategies, which will be similar , due to which concerns how to apply artificial intelligent technology to achieve spacecraft for space exploration mission or space tourism how to raise efficiency and productivity and avoid or reduce space stones collision accident occurrence chances in dangerous space environment.

Hence, space tourism or space exploration mission organizations need to concern above these aspects to implement their strategic plan. So, we know that the strategic plan for artificial intelligence technology to be applied to space exploration missions or space tourism missions which aims to help space operators and space pilots to make predictive judgement to reduce or avoid sudden space accident, e.g. space stones collision to the spacecraft to cause space operators or space pilots hurt, even die as well as how to apply artificial intelligence to assist space operators/space pilots to raise efficiency, when they are carrying on any space exploration missions or space tourism leisure missions.

- How to achieve artificial intelligence technology to space exploration mission or space tourism mission more easier and more effective?

I recommend pace tourism or space exploration mission organizations can set up one examination structure of strategic council for (AI) technology to assist space exploration mission or space tourism development. When the strategic council for (AI) technology was established the (AI) research coordination council and space industry coordination council were also needed to be researched.

The (AI) coordination council needs to be progressed with giving shape to linkages in (AI) research and development carried out by there ministries. The space exploration mission or space tourism industry coordination council needs to carry out surveys and investigations on (1) establishing a roadmap for space exploration or space tourism industrialization, (2) fostering of space operators, space pilots, space tourism leisure customer service operator of human resources, (3) data maintenance/provision and open tools, (4) measures, such as for fostering start-ups and financial linkages, in aiming towards how to apply (AI) technology to assist any space exploration mission or space tourism research and development carried out more easily.

With regard to ethical aspects of (AI) technology to space tourism or space exploration mission, intellectual property right, personal information protection and promotion of open data, separate opportunities for examine whether the (AI) strategic plan is suitable to the space tourism leisure business or space exploration mission business. If the organization feel it is not suitable to the space tourism leisure or space exploration mission business. Then , it can attempt to find what weaknesses are for its (AI) strategic plan and to revise to correct its (AI) strategic plan more easier.

The (AI) strategic plan can concern how to apply (AI) industry roadmap technology to assist to related space exploration or space tourism technologies . Thus, the (AI) strategic plan includes new (AI) automated control system assistance service and space robotic products to be utilized and applied of (AI) technology to spacecraft technology. It aims to avoid or reduce sudden accident occurrences and it can predict when and how and why sudden accident occurrences as well as how to solve these challenges urgently during the spacecraft is flying on the space environment.

Hence, the strategic plan of (AI) technology must assist human space pilots/space operators to raise any space exploration missions or space tourism missions efficiency and it must assist them to predict when sudden accidents will happen, e.g. predicting when and where space rocks collision to the spacecraft in order to avoid the accident occurrences in space environment.

The (AI) strategic plan can include these three phases to any space tourism or space exploration missions in space industry. (AI) technology is simply a service to be provided to space operators/ space pilots to use to assist them to make more efficient and reduce or avoid sudden accident occurrence chances in space environment. The three phases are as below:

(1) The first phase: Utilization and application of data-driven (AI) automated control systems/ space robots developed in the space environment. This phase aims to achieve space (AI) automated control systems and / space robots to raise space service efficiency.

- On real-time assessment of space

On operator operational status hand, it includes such as below:

How to use space robots on any unmanned space tasks assistance to space operators / space pilots? How to predict when failure of space equipment damage occurrence and how to solve those sudden accident occurrence challenges in spacecraft during it is flying gin space environment? How to apply (AI) -based prediction / matching of supply and demand to every spacecraft its space exploration mission or space tourism trip need, such as on-demand space (AI) robotic assistance supply service, optimization control of spacecraft energy consumption by using (AI) automated control energy management system technology, cooperative space tasks by human space operators/space pilots and space robots, implementation of space robots that simulate behavior of space pilots and space pilots, enhancement of space operators / space pilots cooperation with space (AI) robots? Which aims to create of new cooperation services with space (AI) robots / space (AI) automation control systems assistance in order to achieve any space tasks more efficient, reducing / avoiding any sudden space accident occurrence chances from manual or non-predictive space environment factors.

(2) The second phase: Space robots able to perform multiple functions and cooperate with each other to assist space operators or space pilots to do different tasks in the spacecraft, creation of diversified space robotic services.

(3) The third phase: It is the final strategic plan stage. It aims to achieve any space exploration missions / space tourism leisure missions which can achieve innovative space services and space (AI) robots are continuously developed to assist space operators / space pilots to do different tasks efficiently and effectively in shorten time, during they are carrying on any tasks in any spacecraft.

The final phase of space strategic plan objectives include as below:

- Space (AI) robots can assist to space operators / space pilots to solve manual task errors discovery in order to help human space pilots / space operators to solve challenges in any time in spacecraft.
- Space robots that can provide walking assistance, supervision, and support through conversation to cooperate with space pilots / space operators to attempt to solve challenges in any time in

spacecraft.

- Space robots which can understand the space operator / space pilots intentions or needs how to finish or do every task or space mission efficiently and effectively.
- Space robots can collect of space traveler individual information and predict of when space environment will change using (AI) space robots and sensors. For example, space robots can take 3 D maps or photos , when they need to be assisted to take any space exploration mission photos on Mars or Moon etc. planet lands.
- Diversification of spatial space transportation vehicle devices, such as (AI) space vehicle drones assist space operators to deliver cargos / carriages on any planet lands.
- Diversification of space transportation devices, providing valuable space transportation service for space travelers.

Consequently, (AI) space strategic plan aims to let space operators lead space robots / space operation control systems can follow these three phases (stages), in order to have right or reasonable direction to let space operators, space pilots, space tourism service operators, space exploration mission operators who can also follow right or reasonable direction to achieve whose space tourism leisure aim or space exploration mission more easier and efficient from (AI) technology assistance.

- (AI) robotic technology brings long term or short term benefits to space exploration industry

Whether do scientists spend time and money to invest artificial intelligent invention to space exploration industry, is it worth to earn short term benefits only ? As NASA exploration moves beyond earth's orbit, the need excites for long duration space systems that are needed to ensure events that compromise safety and performance. I believe that the (AI) technology advances in autonomy, robotic manipulators, artificial intelligent in-space vehicles can provide service possible at acceptance cost and risk.

I shall explain how to evaluate future space systems needed to support scientific observatories and human/robotic Mar, Moon planets exploration to access key structure design considerations.

How can impact of in-space robots, (AI), in-space vehicles to support NASA's future long duration missions?

(1) Robotic in-space vehicle

In the future, new knowledge is gained by robotic scientific observatories like the Hubble space telescope and the Mars covers will have result to be applied to commercial and government spacecraft launch needs. Why do future space exploration missions need (AI) technological assistance? The reasons include improving aperture size, decreasing deployment risk, such as assembling systems are too large to fit into a single in-space launch vehicle and enabling repair and upgrade easily if any future space exploration missions choose to apply (AI) technological assistance. It will reduce cost and accident occurrence risks.

In this Mars exploration case, NASA's science and exploration missions of the future require spacecraft systems, both robotic and human tended, they can cooperate in deep space for extended provider of time. The Mars exploration mission could used (AI) tools to assist human space operators to test deep space habitation technologies for a Mars transport habitat. The robotic in-space vehicles could also be repurposed as an exploration platform surface exploration and provides a deep space vehicle assembly and servicing site in Mars. So, the robotic in-space vehicles can assist space operators to deliver cargos in Mars, different space sites.

Moreover, the space (AI) in-space robots have also these advantages, such as free-flyer robots which can fly to Mars different space sites to observe Mars different space sites situations to find whether what needs they need to assist, fuel storage function repair and assembly function, long reach manipulators functions power / data / mechanical joining technology function, high power, mass efficiency, large scientific observatories function.

Hence, future robotic in-space vehicle tool which can assist space operators to observe Mars different space sites easily, transport any cargos on Mars different sites between them easily. Even, it can assist space operators to attempt to find whether where has water resource, and find suitable farming land locations grow food to provide human to eat in Mars, and find alive existing locations in possible. Thus, space robotic in-space vehicle is a good example to assist space operators to achieve future Mars exploration mission for long term time.

- Robotic spacecraft

In the future, the first space robots will possible be spacecraft and probes and the main goals of these machines was the Moon or Mars surface aiming manned missions to the Earth natural satellite. The capability of these robotic spacecraft included photographing and recording the surface and sending television images to the Earth.

The first robotic spacecraft will send to the Moon had not even the capability to land in the surface. They were lunar impact robotic spacecraft. Television images were obtained and send to the Earth when falling down on the Moon. Other robotic space achievements by using probes were directed to Nenus, Mars and Mercury, Jupiter and Saturn.

However, space robotic spaceship will face these challenges to be solve if space scientists expect future space robotic spaceships can attempt to fly to Moon, even other planets to reach these any one of space planet destinations successfully.

The challenges include that it supposes the space journey needs to reach the space region with Voyager 1 and the most distant planets with Voyager 2. The robots are needed to be the appropriate machines to substitute the astronauts in risky activities in the non-predictive dangerous space environment. How to design the artificial intelligent robot will be more important and will be a key issue for the future of space exploration?

The orbital robot requires to solve these problems , such as radiation, strong temperature variation, micro-gravity environment and magnetic fields associated to some celestial bodies. Specially, for planetary exploration aspect of the robotic application includes surface exploration where the local gravity, soil characteristics, local pressure and atmosphere must be consideration. So, future space robotic tool development still needs to find solutions to deal these above challenges when it brings to space environment to be used.

- Space navigation and long distant mobility

In the future space stations need have robots to help them to repair when sudden accident occurrences to cause any one of space stations' machines to be damaged. In-space operations involve operations functionalities like in-space assembly, in-space inspection, in-space maintenance, extra-vehicular activities, and in-orbit scientific experiment. For example, planetary exploration robotic applications are remarkably operating on Mar surface.

Space scientists predict the next 10 years, it is expected that the space robots for surface exploration be not constrained anymore by, navigation and long distance mobility to access to most locations on a planetary surface will be possible. On the other side, it is expected that ground based planning and visualization tools enable scientists from space ground station to interact with the space robots in the surface of celestial bodies. However, robotic performance at the level of a space suited human scientist is and will continue to be a major challenge.

Anyway, the automation by using robots in space operation will bring the advent of space stations and orbital service. One of the most important accomplishment with an extraordinary space robot was the space shuttle series of spacecraft. Space shuttle can be considered future first space robotic shuttle can be provided series of spacecraft service. Thus, future space robotic shuttle can be considered in that if really implemented orbital servicing of grasping satellite for maintenance, and assembly for the international space station and inserting new satellite into low Earth orbit.

Consequently, robotic technology can bring long term benefits to any space exploration missions. It is absolute short term benefits. For example, robotic vehicles can assist space operators to deliver cargos in long distance between different space stations fast and easily. They don't need manual control. Such as on Mars or Moon planets which will let many robotic vehicles move on their lands to

move lot of building materials to build houses or space stations or space warehouses or space laboratories on Mars between different space sites. The benefits are that they do not need space operators drive them to more on Mars lands. They can move on Mars land automatically. So, space operators can concentrate on doing their space exploration tasks.
For another example, space robots can take 3D photo images in Mars easily. They can deliver 3D photo images more clear and visible in short time from Mars on Moon planet to Earth space stations. They do not need space operators to control, they can automate to take 3D clear photo of Mars or Moon current images more clear in order to let Earth space station scientists to see these visible 3D Mars or Moon planets images to gather data to prepare to do any space exploration missions immediately as soon as possible.
For final example, future space robots can learn space operators to attempt to assist them to do any maintenance tasks. So, when the aircraft encounters sudden accident to cause machine (equipment) damage, then the space robots can assist to the space operators to find the damaged equipment to repair immediately. So, they can share space operators' maintenance tasks risks if human operators feel difficult to find which are the damaged equipment which need to be repaired. Then, space robots has possible to find which are the damaged equipment to be repaired immediately. Hence, future space robotic technology seems to bring long term benefits to any exploration missions absolutely.

● Future space tourism psychology
prediction stragegy

● Psychology and economic environment changing both factors influence whole space tourism market leisure desire

How can apply (AI) technological and psychology method to attract future space tourism customer individual space tourism desire? I believe that it has relationship between the space tourism planner and the safe (AI) space tourism environment to attract every one

space tourism passenger to catch the (AI) space boat to fly to space to travel as below:
Firstly,the space travelling planner individual psychology influence hand, the space travelling planner will have these both aspects of individual psychological influence, it includes these both either positive or negative psychological influence aspectsas below:
On the positive psychological influence aspect, if the space travelling planner has confidence to the space travelling leisure company can provide safe, comfortable, good quality of one space travelling trip arrangement, good taste food arrangement, reasonable space ticket price and every reasonable space trip for space hotel living arrangement and space garden and space farming land visiting journey arrangement, even, space swimming pool and space sport centre and space cinema leisure arrangement to let whom to stay on the planet at least one day trip, it means not one short time space trip, e.g. the spacecraft only flies about half hour or one half. It can not fly to the planet to arrive its space station destination to stay to let the space travelling planner to live at the space hotel at least one night. Then the space travelling planner will have more desire to choose to catch the space tourism leisure company's spacecraft to travel to space.
Secondly, on the negative psychological influence aspect, if the space travelling planner lacks confidence to the space travelling leisure company can provide safe, comfortable, good quality of one space travelling trip arrangement, good taste food arrangement, reasonable space ticket price and every reasonable space trip for space hotel living arrangement and space garden and space farming land visiting journey arrangement, even, space swimming pool and space sport centre and space cinema leisure arrangement to let whom to stay on the planet at least one day trip, it means not one short time space trip, e.g. the spacecraft only flies about half hour or one half. It can not fly to the planet to arrive its space station destination to stay to let the space travelling planner to live at the space hotel at least one night. Then the space travelling planner will have less desire to choose to catch the space tourism leisure

company's spacecraft to travel to space.
Hence, it seems safe (AI) technological space boat factor and the space travelling planner's confidence factor to the space tourism leisure providers will influence the whole space travelling market whose space travelling consumer's space travelling leisure consumption desire to be more or less. So, any one space tourism provider will need to consider how to apply (AI) technology to assist space boat safety how to influence whose customer consumption desire.

- space tourism strategy

Future any space tourism leisure business needs have good business plan to outline the space tourism leisure business in these aspects , such as: different space tourism destinations of every space tourism journey, technical , financial and regulatory factors for growing space tourism leisure consumption into any one kind of unique artificial intelligent space tourism journey for identified passenger target group.

All how to design one space tourism business development plan to attempt to predict whether what trends will influence how every different kinds of identified space tourism journey in order to achieve passenger number growing aim as well as how to achieve one attractive space tourism leisure to satisfy future space tourism passenger individual space travel needs more easily.
I shall indicate what aspects to future every space tourism traveler who will consider in order to reduce the space tourism traveler personal worry to catch any pace boats to leave our Earth to fly to other planets to travel.
I recommend that any space tourism leisure organizations need to concern how to to apply (AI) technology to these aspects in their space tourism leisure business plan as below:

(1) safe space tourism journey

On first aspect concerns safe space tourism journey plan to let all space tourism travelers will considerate safe issue. They must ensure space boats that is safe to catch them to fly to planets in their space journeys. So, any space tourism leisure business will utilize previous flight rated and proven technologies to form the basis for manufacturing spacecraft vehicles, and will incorporate the latest modern avionics and flight systems for ensuring safety, reliability and economical operation in order to reduce any space tourism traveler personal worry to catch any spacecraft.

So, the space tourism safe journey plan is one very important factor to influence space tourism consumer number for them if any one of space tourism leisure business hoped they can grow the space tourism consumer number for long term. For example, the space boat flight hardware must often be maintained at the space station. It is needed to be considered by space boat experts as risky, extremely expensive and potentially sensitive. To aims to ensure spacecraft will offer an economical and safe alternative for any satellite manufacturers and other space tourism entertainment organizations have a desire or requirement for space tourism flight.

(2) reduction cost expense plan

On second aspect concerns reduction cost expense plan, any space tourism entertainment organizations need have the experience and capacity for safely launching a fully loaded , including space tourism passengers and passenger individual cargo for every spacecraft tourism journey. As a result of outsourcing the launch role to a major contractor, the space tourism pilot can concentrate on space boat crews flight training, planning space tourism passenger cargo capacity and preparing space flight manifests , and will as a result, avoid the expense of maintaining a launch operation on a daily basis.

In addition, by outsourcing the spacecraft manufacturing, it can avoid spending millions of dollar on facilities and equipment infrastructure and engineering manufacturing expertise.

(3) achieve any space tourism mission plan

On third aspect concerns how to achieve any space tourism mission. Every space tourism mission must be ensure that reliable service is provided to satisfy every space tourism passenger personal space traveler needs and let them to enjoy in their whole space tourism journey, let them to catch a big aircraft in comfortable environment of technologically sophisticated space boat, reasonable and competitive every time space tourism flight ticket price plan is developed and properly revised every time space tourism ticket price when performing their assigned every different space tourism journey mission.

Hence, the space tourism leisure company will provide one careful selected space tourism destination , e.g. Mar planet space tourism journey, Moon planet space tourism journey or no any space destination journey, it means that the space craft only needs to fly one circle around between Earth and Moon space journey etc. that are capable of meeting the requirements of travelling into Earth orbit. So, any space tourism journey must emphasize affordability, reliability, safety, customer service and responsiveness in responding to every client's space tourism journey requirements. Hence, any one of space tourism journey must have clear space journey mission and objective to satisfy any space traveler client target needs.

- Methods to raise space traveler number

Future space tourism will be one kind of new travel leisure market for any new space travel leisure companies to enter this undiscovered market in the beginning. However, how to predict future 10 to 20 years , even more space traveler number that is one important issue to any new space tourism leisure companies.

I think that space tourism leisure companies need to apply (AI) technology to define what kinds of space travel leisure service to be provided to space travelling passengers, however, what age group of space passengers who will be their space travelling target client. For example, their space travel leisure must provide any flight operation that takes one or more passengers beyond the altitude of 100 km and thus into space to let space travelling passengers who have fun, exciting space travelling feeling.

Anyway, for any kind of space tourism (leisure space travel) journey, space tourism leisure company needs anyone to be bring customer satisfaction, it is a plan or predictive methods to measure how to let every space travelling passenger to feel comfortable when they are catching the spacecraft (space flying product) and they can have enjoyable and fun or exciting feeling when they have need providing any space tourism journey, services meet or surpass customer expectations.

Thus, any space tourism leisure company needs to evaluate the degree of every time space tourism journey's customer satisfaction and customer satisfaction is also always evaluated in relationship to the every time ticket price of the space tourism journey. So, the space tourism leisure company will predict the next time of what the space tourism journey of passenger number is more accurate, after it has evaluated what degree of every time space tourism journey's customer satisfaction is. It aims to gather their opinions to find which aspects that they need to revise, e.g. choosing where will be the next time space tourism journey destination, how to improve spacecraft staff's service attitude and performance to serve to their space tourism passengers when they are catching the spacecraft, to evaluate whether the spacecraft can provide comfortable and safe environment to let them to catch in order to let the next time space travelling passengers can feel satisfactory and enjoyable when they are catching the space tourism leisure's spacecraft to fly to anywhere in space.

In general, the expectation of factors space passengers include the following customer value elements, such as below:

- viewing space and the Earth.
- experiencing weightlessness and being able to float freely in zero gravity.
- experiencing pre-flight astronaut training and related sensations.
- communicating from space to significant others.
- being able to discuss the adventure in an informed way.
- having astronaut like documentation and memorabilia

These objectives need to be combined with, sometimes conflicting constraints, such as guaranteed safe return, limited training time, reasonable comfort, and minimum medical restrictions. All these above issues which will be every space travelling passenger considerate matters before they choose the space tourism leisure company to catch its spacecraft to fly to space. So, all these factors will influence the next time space passenger number. Any space tourism leisure company can not neglect how to solve these all matters before they decide when their next time space tourism journey to be achieved.
Consequently, if the space craft tourism leisure company could revise what aspects of its last space tourism journey to find what are its wrong or weakness or unattractive challenges to cause any one space travelling passenger who feels unsatisfactory. Then, it can have more effort to concentrate on improving its next space tourism journey to raise its space tourism service performance level , e.g. people, food, leisure etc. service aspects and its space tourism product quality level, e.g. proving comfortable spacecraft facilities to let space travelling passengers to catch in whole spacecraft tourism journey. Then, it will have more confidence to achieve the raising space travelling passenger number when any space tourism entetainment providers can apply (AI) technology to assist their space tourism mission development.

● What is the prediction space travelling passenger desire method ?

The prediction space travelling passenger individual desire method can be one survey investigation method. When every time spacecraft finishes space tourism journey mission, after all space tourism passengers catch the spacecraft to arrive earth from space. When they arrive earth space station destination, then the space tourism leisure company can arrange survey investigation staffs to enquire their feeling for this time space tourism journey immediately.
The survey content can include as below:

Do you feel satisfactory or unsatisfactory to which aspects of this time space tourism journey?
(1) On service aspect questions include as below:
(a) Do you feel space food taste is good?
(b) Do you enjoy this time space tourism journey arrangement?
(c) Do you feel satisfactory to space staff
service performance?
(d) If you have unsatisfactory feeling for any one of above questions, which aspect issue cause you feel unsatisfactory to explain to let us to know in order to us to revise our service performance.

(2) On product aspect questions include as below:
(a) Do you feel comfortable when you are catching our spacecraft in whole space tourism journey?
(b) If you feel comfortable , may you explain the reasons what aspects of our spacecraft has weakness to cause you feel uncomfortable?
(c) Do you feel safe when you are catching our spacecraft in whole space tourism journey?
(d) If you feel unsafe, may you explain the reasons what aspects of our spacecraft has weakness to cause you feel unsafe?

Finally, we thank your ideas to be given to let us know how to improve our every time future space tourism journey in order to find what challenge cause our service performance and product quality which can not satisfy your needs. So, we shall improve to avoid future challenges continue occur.
Our mission is achievement of 100% satisfactory level to our every space travelling passenger individual feeling. Also, we hope that you can choose our space tourism leisure service again, when you have another time space tourism leisure desire need. However, we shall revise to improve our service performance and product quality to be better, after collecting your ideas from this time survey investigation. I think you spend time to give your ideas from this survey investigation faithfully.
So, survey investigation method will be one important idea

gathering tool to help any space tourism leisure company to revise the weaknesses to raise or improve future every time space tourism journey service performance and product quality to achieve raising competitive effort in this new space tourism leisure market.
Hence, survey investigation method will be the best idea gathering method to predict how space travelling passenger emotion or desire need will change in order to achieve the objective of raising every time space tourism journey future space travelling passenger number more easily for every space tourism leisure company.

- The prediction of price factor influences space traveler number

The space tourism leisure organizations indicate the total cost of a trip into space is rapidly coming down from the initial price level of about US$600,000, it is obvious that the space travelling customer base is going to be rather small. Typical customers tend to belong to the top 1% income bracket. They also indicate that the price comes down , it is expected that new space travelling customer groups will enter the space tourism leisure market.
Typical new customers include people in other brackets with one-of-a kind incomes, such as inheritance or business sold. There are indications that those types of customers are becoming interested in spending on an once-in-a lifetime space experience. Therefore, the growth of the space tourism market is highly sensitive to customer satisfaction and how it is communicated through various media.
This will establish the status factors of space tourism and corresponding brand reputation service providers. They also suggest that any operator monitors space travelling customer satisfaction closely, as it will help developing increasingly accurate estimates of how the space tourism leisure market will develop.
Hence, it seems that every time space tourism journey price variable factor will influence the time space tourism of customer individual leisure desire and the space tourism passenger number. For example, the minimum price goal foe a variable space tourism

business is currently estimate to be below US$3000-4000/kg for a round -trip depending on variable configuration and operation size. At this price, they estimate that somewhat over 1 % of the high income earners are potential customers.
However, for significant volume growth the longer term goal should be below US$2000/kg for a typical passenger, baggage and supplies. The lower price will probably open space tourism to a broader population, expanding the customer base and altering expectations. beyond this point space tourism will become into a travelling competitive leisure commodity, price competition will ensure and service providers need to rethink their space tourism marketing and branding and price strategies.
I shall also recommend how to attract the potential customers successfully. First, space operators need to pay special attention to the right level of customer services. Second, various preparatory customer operations cost, such as a travel to the launch site, space tourism destination accommodation, pre-flight training, medical check-ups and equipment my add up to between 10 to 15 % of the actual space travel cost. Thurs, solving the right balance between services offered and cost of client operation in order to earn the largest intangible benefits, such as loyalty, confidence, leisure enjoyment, comfortable space travelling journey as well as tangible benefits, such as profit, spacecraft manufacturing facilities, space stations, space hotels , space swimming pools, space gardens, space cinema etc. which are built to similar to earth building facilities to satisfy space travelers' needs.

- The influential factors persuade travelers choose space tourism

Nowadays, our earth is no longer an adventurous enough place for some experienced tourists. Space tourism will be a new sector of adventure tourism, which is in the near future will be fast becoming a new tourism leisure opportunity for experiencing the unknown. Of one day, space tourism is able to reach the mass tourism phase, due to improved safety and decreased operation costs, a future space tourist will possibly only need minimal training to cope with

the zero cost.
Space tourism is quite well established with visits to space attraction and launch sites, and it is a wealthy trips to the international space station for any space tourism travelers. However, if any space tourism leisure companies can attempt to find what the most influential factors are to persuade travelers feel attraction more than travelling in our earth.

It aims to let travelers to choose space travelling more than earth travelling when they feel travelling leisure need. I shall indicate what will be the most important influential factors to persuade travelers to choose space tourism more than earth tourism as below:
Firstly, I shall argue that the majority of different new space tourism journey destinations will be needed to find to satisfy different aged space travelers and different income space tourism consumers' needs. For example, the rich people have effort to consume longer time and reach any space tourism destinations where are far away from our earth of their every space tourism journey.
Otherwise, the middle income people will choose shorter space tourism journey distance from our earth and short time space tourism journey. Also, younger space tourism clients can accept more longer journey time, exciting fast speed spacecraft flying journey. Otherwise, old space tourism clients can only accept comfortable and shorter time safe space journey. So, it seems that safety, comfortable feeling, shorter time space tourism journey won't be one important influential factor to excite any young people who choose to consume space tourism leisure. Otherwise, safety, comfortable feeling, shorter time space tourism journey will be one important influential factor to excite any old people who choose to consume space tourism leisure.
Secondly, the another most important influential factor to excite space travelers to choose space tourism , it concerns whether the space travelers will feel what tourists benefits can be earned from a substantial variety of destinations choice. In general, space tourism with those of aviation, space travelers will hope space tourism will be travelling distances by air in a very short time, safely and

comfortably, to bring them to arrive any space planet destinations when spacecraft reaches any space stations to stay in any space destinations.

Hence, space destination factor will bring important influential choice to any space destination journeys. As a result of the space technological tourism boom, the number of potential different space destination, choice attractions have grown with far fewer places on earth to which human do have access yet. However, the ultimate different space destinations to which many of us dream is not on earth, but as least 100 km above us, anywhere in space any planets.

If the space tourism leisure company can provide different space tourism destination choices to young or old age both space traveler target consumer groups. They will feel a real holiday when they will be able to enjoy a great image of the earth from planets. It might mean that every space tourism journey can provide different space tourism destination to let space travelers have another new travelling destinations where are far from our earth anywhere.

Hence, the different space tourism destinations will give them an unforgettable adventure. Think of how it would be to be able to check in at a " billion strategy" luxury hotel in space one planet, it means that the space planet destination can provide one luxury hotel to let space travelers to live one night or more in the space planet destination, how it would be to schedule the space traveler' vacation at one of the space tourism leisure company luxury resorts on the Moon or Mars.

This images seem from science fiction movies, but one should not forget that 100 years ago, the Wright brothers, aviation pioneers inventors and builders of the air plane, would not have imagined how, every day it is possible that future spacecraft can fly to any planets to let human have chance to stay in the space hotel one night or more.

Consequently, space destination choice and space tourism journey service performance, aviation safety, ticket price and leisure satisfactory feeling which will be important influential factors to attract future space travelers to choose space tourism leisure to replace earth tourism leisure in future one day.

- Raising space tourism leisure
consumption strategies

Although, space tourism industry is a real enjoyment and exciting travelling leisure to human. It is possible that human will choose to consume space tourism leisure to replace earth tourism leisure, if human felt that earth tourism leisure is not attractive to them to consume to go to anywhere to travel in their leisure time.
But, I believe that space tourism industry has still many factors to influence human to choose to consume space tourism leisure, even they will consider space tourism leisure consumption I is only one time space tourism in their life time. Hence, space tourism companies ought achieve this aim to persuade or attract everyone prefer to spend space tourism leisure at least one time in their life, then it can represent success. However, I think to achieve this aim, it has these challenges to influence their success, even they believe space tourism leisure business is one potential attractive travel entertainment business. These challenges include such as: expensive space tourism ticket price issue, catching spacecraft safe issue, space traveler personal body health issue, age issue, family and friend relationship influence issue, working time and holiday time arrangement issue, the space trip arrangement issue, weather issue etc. different challenges, which will have possible to influence every space tourism planner either who decide change to cancel the time space tourism plan, or forgive to choose space tourism leisure in their life forever.
Hence, how to raise space tourism leisure consumption desire will be one considerable matter for any space tourism leisure businessmen. I shall indicate my personal three aspect of strategical opinions to let them to know how to raise every space tourism planner individual space tourism leisure consumption desire to avoid every time space tourism passenger number will have decrease failure chance as below:

● (1) Strategic opinion

On the first aspect of strategic opinion, I feel that the space education tutor can teach new space knowledge to let every space traveler to learn any new space and earth knowledge during he/she is catching on the spacecraft in personal contact learning experience environment which can raise space tourism consumption desire. The reason is because the space tourism leisure traveler can raise extra space and earth learning knowledge when they can catch the spacecraft to fly and contact the space environment to learn and feel what the differences are between space and earth by himself or herself. Hence, it is very attractive to the space traveler student target group and I believe that their parents will encourage their sons or daughters to participate the time of space trip and they are more preferable to help them to buy the time space trip ticket, due to their sons and daughters can learn any space knowledge when they are studying. Moreover, every space traveler will feel surprise to learn any new space and earth knowledge from the space tutor's teaching, due to he/she is unknown that this space travel trip includes learning space and earth knowledge.

I suggest that the space tourism leisure businessmen can give learning opportunity to every travel trip space travelers to feel that this space actual environment can bring what disadvantages or advantages to influence our earth when they are catching aircraft to fly to space to travel in every space trip. The space and earth learning knowledge can include these two aspects of space learning knowledge and experience below:

On the teaching of space environment learning knowledge hand, the topics can include as below:

Firstly the space learning topic can concern how space environment influences water and hydrated minerals change , they can learn what our drinking water function how is applied to space environment. For example, in the space environment, they can learn and attempt to feel that how water can be used in protecting

astronauts against harmful radiation from the sun and cosmic rays by cloaking spacecraft with a thin layer of water in the actual space environment as well as the space travelers can also feel water is same as fuel when they are catching the spacecraft, they can feel the water is heavy to transport into space when they are catching the spacecraft to fly to space during their whole space tourism journey.
Moreover, when their spacecraft reaches anyone of planets and it stays on the planet's space station, e.g. Moon space station. They can learn how to attempt to contact the hydrated minerals to learn and feel what they contained in some asteroids may be possible sources of water and fuel in the actual space environment. When they are walking in actual space environment, such as Moon planet, they can contact or touch this hydrated minerals to learn how water molecules can be extracted and separated chemically to produce hydrogen fuel knowledge in the actual space environment. This is one exciting space learning experience to the space travelling student passengers.
Secondly the space learning topic can concern how human fights space threats , even when their whole space leisure journey, the space science teacher can let the space trip student passengers to feel that they are learning new space knowledge between the space science teacher and whose space trip student passengers. Such as how to protect our earth knowledge: Teaching them to know when will be threats to our earth from space. The space science teacher can explain how this space threating environment influences our life safety and let them to feel that a mass extinction can be triggered if an asteroid 10 kilometers across hit the earth. Even being the apex species in the food chain did not space carnivorous dinosaurs from such disaster, who knows if this terrifying scene won't happen before our eyes? So, the space travelers can image and feel how the space threating environment can influence their life safety in the actual space environment as well as the space science teacher can let whose space travelers to feel and image the actual earth disaster will possible happen suddenly to let they feel afraid in the actual space environment. Also the space science teacher

can teach how our earth can fright the space stones attack to let the space traveler to know, when an impactor targets an asteroid for a controlled well-times wallop. The collision will change the asteroid's momentum, deflecting it from its original orbital path which intersects with that of the earth. So, at the moment, the space travelers can image they are a larger spacecraft near an asteroid which can also change the path. Given enough time, the gravitational pull from the spacecraft will be able to steer the asteroid away from the earth. So, every space traveler will feel that they are catching the spacecraft in the safe space environment to avoid the Earth disaster from space sudden unpredictable attack.
It is more fun real space tourism knowledge learning feel to let every space traveler has chance to learn any new space science knowledge when he/she is catching the spacecraft to fly to space to travel. Hence, one successful space trip ought include trip and learning experience both contents in order to raise every the space tourism planner individual space trip consumption desire.

● (2) Strategic opinion
On the second aspect of strategic opinion, space tourism leisure companies need to let planning travelers feel that anyone of space tourism leisure is very different to general tourism leisure. In general, tourism leisure is visiting at least one night for leisure and holiday, business or other tourism purposes in Earth only. Otherwise, space tourism leisure is other kind of an unique trip leisure or entertainment method, e.g. the space traveler can catch the spacecraft to visit any planets to stay to live at the planet's space hotel at least one night, e.g. Future potential populated Moon or Mars space hotel space trip. Moreover, the space travel companies ought give chance to let them to feel what weightless feeling is in weightlessness environment when they are walking on Moon or other planets in possible. Even, they can attempt to build these entertainment facilities, instead of space hotels, such as space swimming pools, space gardens, space cinema etc. building facilities. It aims to let them to feel what the differences between

Earth and space life when they are walking on the Moon, when they are swimming on the space pools, when they are living in space hotels, when they are watching movies in space cinemas, when they are seeing flowers and different species of planets and fruits. e.g. oranges, apples, bananas, and vegetable and potatoes and tomatoes in space gardens. It is very exciting and fun space trip life experience between one days to seven days. So, they believe that they must not feel these space life experience if they do not choose to participate this time space trip planning journey by the space trip company preparation.

Also, due to that the space tourism passengers need to the pre-flight checks and training before they ensure to qualify to permit to participate the space trip. So space travel companies need to concern how to take care their health check and training matter considerately. It aims to let every space traveler will feel a market segment with fitness and extreme experiences as well as he/she will become popular with a market segment passenger to the space tourism leisure company, although he/she must not guarantee to pass the space training and/or pre-flight health checks to permit to participate the space trip. However, he/she can believe that he/she is one worth space travelling passenger to the space tourism leisure company, even this time pre-flight health check or/and the short time space trip training requirements are failure. However, the space tourism leisure company must need to let all pre-flight health check and space trip training passengers to feel that it is only one space tourism which can give them and let customers view the space travel is as the ultimate showcase for health, even though a majority of the population can pass the pre-flight medical and other tests in order to raise their confidence and safety to catch the spacecraft to fly to space to travel when they are confirmed to pass these tests to permit to catch the spacecraft later.

In general, the expectations of future space passengers include the following customer value elements, such as below:

- Viewing space and the Earth.
- Experiencing weightlessness and experiencing pre-flight

astronaut training and related sensations.

- Communicating from space to significant others.
- Being able to discuss the adventure in an informed way.
- Having astronaut-like documentation and memorabilia.
- Enjoying one exciting and fun space trip.

However, instead of considering these objectives need to be combined with, sometimes conflicting , constraints such as guaranteed safe, return , limited training time, reasonable comfort, and minimum medical restrictions. So, space tourism companies need to reduce every space traveler individual worries before they decide to make the time of space tourism journey. Then, it can increase their confidence to raise their space tourism consumption desire more successfully.

Consequently, instead of these consideration, a space travel operator must pay attention to the total customer experience over the entire customer process, starting from how the service is presented, proposed and sold. The service package must include training, instructions, travel to the launch site and various post. Travel activities to generate maximum customer satisfaction and brand building opportunity.

- (3) Strategic opinion

On the final aspect of strategic opinion, I think any space tourism companies space tourism companies need to consider every time space tourism ticket price and space tourism trip issues. It is important factor to influence every space traveler individual consumption desire. Due to space trip ticket price must be more expensive to compare common Earth trip travelling ticket price, so this kind of tourism leisure market target customer will be the rich and high income customer group.

On the space trip ticket challenge issue, despite that fact the total cost of a trip into space is rapidly coming down from the initial price level of about US$60,000, it is obvious that the customer base is going to be rather small and the client target customer is only high income or rich consumer group. Typical customers tend to belong to the top of the top 1% income bracket. So, ensures that space traveler

number must be less than common Earth traveler number.
Also, such as the space trip ticket price, it is expected that new middle rich level or middle high level income customer target group will enter the space trip leisure market, when every space trip ticket price falls down about 1% Typical new customers include people in other income brackets with one-of-a-kind incomes, such as inheritance or business sold space traveler target group. These people will be space travel new client group, when its every space trip ticket price can be reduced to close 1 to 2 % nearly. If any space tourism leisure companies expect to attract new rich and/or high income target customer group to choose any one kind of space trip journey planning to consume.
These are indications that these types of customers are becoming interested in spending on an once-in-a-lifetime space experience. Therefore, the growth of the space tourism market is highly sensitive to customer satisfaction and how it is communicated through the various media. This will establish the status –factor of space tourism, and corresponding brand reputation of service providers. The minimum price goal for a variable space tourism business is currently estimate to be below US$3-4000/kg for a round-trip depending on vehicle configuration. So, space travel leisure companies need to concern every round space trip cost, it can depend on the space vehicle number and weight issue to influence every space trip ticket price variable to achieve how much it can earn.
On space journey design factor aspect, it includes these different facilities aspects how to design, because future space travelling consumers will concern whether the space travel company can provide special entertainment to satisfy their needs. The facilities include as below:
How to design space hotels to let them to live in comfortable space environment and eat the best taste and fresh food quality when the cookers need to cook in the space hotel in the space environment? How to design space swimming pools to let them to swim in safe space environment? How to design space sport centers to let them

to run more easily in one space sport warm and safe environment? How to design one space garden to let them to see different species of Earth flowers, or plants? How to design one space farming land to let them to see different species of Earth fruits, vegetables, tomatoes, potatoes etc. fresh foods growth in warm and safe space farming land environment? How to design one space cinema to let them to watch movies in one safe and warm space cinema environment? All these facilities will be any one of future space trip's' important and attractive space trip leisure facilities to influence every space traveler to choose to buy the space tourism leisure company's space trip leisure service.

Instead of these space building entertainment facilities, they also need to concern how the space vehicle entertainment tools are provided the entertainment service to satisfy their needs. When the space travelers can sit on the space vehicles to move on any planets' lands, such as Moon. A number of space vehicle options exist in the market, mainly differing based on the seat capacity as well as the in-flight experience level offered. The typical space vehicle solution is a small, relatively light weight spacecraft taking between 2 to 10 passengers. The number of passengers depends on the service level, amenities and extra offered. The trip typically lasts about 10 hours and of which about 4 hours are spent in space. The main attraction is the weightless time after in space. The main attraction is the weightless time after re-entry has started. It is a rather low-G technology and therefore the medical requirements for participants are nor very high.

Consequently, the space vehicles, space leisure building facilities, the space trip reasonable price ticket level, every safe space trip journey arrangement, clean and fresh and good taste space food arrangement, space traveler individual real learning experience etc. these factors will be the main influential factors to raise the space tourism leisure company's competitive effort and the space traveler consumer individual consumption desire to the space tourism leisure company in the future.

- Space travel markting strategy

Any space travel organization needs have good marketing strategy to prepare how to operate its space travelling leisure business in order to attract many space travelling clients to choose its space travelling service. I shall indicate these different strategies aspects whey they are needed to be concerned as below:

(1) On concept of spacecraft design aspect

Firstly, on concept aspect, any one space travelling leisure company needs have at least one spacecraft to catch clients to fly to space to travel. So how to design the spacecraft and its quality and safety and comfortable environment spacecraft machine concept aspect issue which is one challenge to be concerned. Because many space travelling passengers ususally concern whether the spacecraft is safe, comfortable , good quality, as well as the space travelling leisure providers also need to concern whether the spacecraft is less time and energy saving efficient use, less manufactory operating cost and durable.

In general, space travelling leisure provider expects the spacecraft or spacecraft vehicle can be uesed long time. The spacecraft will be expected to utilize previous flight rated and proven technologies to from the basis for manufacturing spacecraft vehicles , and will incorporate the latest modern avionics and flight system for answering safety, reliability and economical operation.

In general, the spacecraft will be designed to carry two crew and approximately, 10,000 pounds of cargo, depending on the ultimate weight of the spacecraft. Relying on flight hardware to maintain the space station, such as Moon or Mar space station is fpr any space travelling spacecrafts to reach these space travelling destinations to stay, it is also need to consider by many space travelling experts as risky, extremely, expensive cost sensitive for any space station travelling destination design arrangement in order to future every spacecraft can fly to any planets to stay on its space station safely.

- Outsourcing spacecraft concept design strategy

As a result, outsourcing strategy is one good method to help them to reduce cost in order to achieve to let every space travel passenger has safe space journey experience and capacity for safely launching a fully loaded (including crew and cargo). Outsourcing strategy is the launch role to a major contracor, they can concentrate on crew flight training, planning all passnegers and cargo capacoty, and preparing flight manifests, and will as a result, avoid the expense of maintaining a launch operation on a daily basis. In addition, by outsoucing the spacecraft manufacturing, the space travelling provider can avoid spending millions of dollars on facilities and equipment infrastructure and engineering manufacturing expertise.

(2) On deciding misson aspect

Secondly, on mission aspect, any space travelling journey needs have a clear mission to be planned how to achieve in order to ensure every space travelling passenger feel satisfactory in the space travelling journey. So, every whole space travelling journey arrangement, e.g. where will be the space travelling destination, how to check every space travelling planned passengers‘ bodies whether who are health to catch spacecraft to fly to space to travel or how to train every space travelling planned passenger to ensure whom can permit to catch spacecraft to fly to space to travel, how to arrange every space travelling journey entertainment and facilities to let either young or old age target passenger to enjoy the space trip to feel satisfactory, how to arrange different days of every space trip. In the last few years, Virgin Galactic has been making new's headlines with its promises to provide space travel services, and announcement that it will soom offer, at quite a hefty price, trips to sub-orbit. It is generally agreed that sub-orbit exists 100 kilometres above the earth's sea-level (Von Der Dunk, 2012). Hence, Virgin Galactic will provide travel to where customers may experience weightlessness, as well as the sight of earth's curvature. Even more interesting is that Virgin Galactic is not the only company with such a mission,there are a few more that wish to offer the same type of service. For example, some companies even aim to provide an orbital type of flight.

Orbit flight suggests that humans would venture into outer space, where they might either orbit the earth or board the international space station (hereinafter: ISS). In addition, some envision space hotels, moon visitations and mining asteroids. Although at first such statement might seem for one must point out that a "space hotel" is already in earth's orbit and that diligent progress through flight tests is almost made the commercial aspect of regular space travel a reality; it is only the question of time and readiness for the companies to make their long-awaited and open a new industry of present day economics (Klemm & Markkanen, 2011; Berry , 2012).

So, every space travelling mission is to ensure that reliable, technologically-sophisicated competitively-priced flight certified spacecraft are designed and properly maintained when performing their every assigned space travelling journey mission. The space traveller leisure provider will need to provide a carefully selected array of techologies that are capable of meeting the requirements of travelling into earth orbit. It will emphasize affordability, reliability, safety, customer service and responsiveness in responding to customer's space travelling requirements.

For this space tourism leisure mission example, it many include these objectives , such as below:

One trip into space, sending a space vehicle of a certain make and with a specify capacity on a space mission, provides the various grades of a core service, such as a space mission including issues such as waiting and delivery times, personal attention and advice, amenities and facilities, ensure quality assurance, it is the planned and system activities implemented in a quality system. So that quality requirements for a product or service will be fulfilled. It aims at preventing high-risk adverse events, or reducing thei impact, provides excellent customer satisfaction, it is a measure of how products and services meet the space travelling customer expectations, customer satisfaction is also always evaluated in relationship of every space travelling ticket price of the space travelling entertainment service and spacecraft product comfortable environment feeling and good leisure arrangement for every space travelling leisure journey.

(3) On space tourism leisure organization managment aspect

Thirdly, on space tourism leisure organization management aspect, it is also important to influence efficient and excellent space service performance to be provided to satisfy every space travel organization management team needs to be consists of experienced professionals who have successfully management and operated companies specializing in the aerospace industry for a number of years.

Their knowledge and contacts within the space industry will prove invaluable in assisting the space tourism leisure provider in the achievement of its goals and objectives. In individuals on the team components that up a spacecraft tourism development organization, and have unique experience in the design, construction, operations and maintenance of the major functions will developing spacecraft for launching into orbit. Every spacecraft will be built and maintained utilizing the same high standards of quality, within budget and well within time constraints.

Hence, every space tourism provider needs have one excellent management leaders to manage every space tourism service staffs to serve passengers in order to achieve excellent service performance to let them every one to feel satisfactory, during their every space tourism journey (trip).

(4) On target audience prediction aspect

On target audience prediction aspect, every space trip needs have identifies target travelling passenger in order to concentrate to choose the most popular and satisfactory space travelling journey for their identified needs.

For primary audiences example, it can include space enthusiasts and educational families both. Space enthusiasts target are usually young people and they are only 20% over 65 age old people target space ethusiasts who will be the future potential space tourism target consumers as well as the educational families target who will aspect owning educational experience for children , who is the explicit reason to visit space, either he/she has interest in history of space exploration or he/she has interest in future of space exploration or he/she feels that spce trip looked like fun.

KSCVC Visots (2013) indicated that future top markets, ranked by high visitation against space enthusiasts and educational families space tourism passengers, the US cities will include: Orlando, NYC, Miami, Tampa Bay, Chicago, West plam, Philadelphia, Atlanta, Boston, Washington, DC and San Francisco cities. So, future US space travelling market will be the top one in the world.

(5) On space objective aspect

On space objective aspect, instead of any one space tourism leisure organization concerns how to achieve its mission to satisfy all space tourism passengers leisure needs. Although, it is the major missin for space tourism leisure industry. But they can not neglect what the objectives are in order to develop or achieve long term space tourism leisure missions more easily.

The objectives main open space key issues can include such as: Providing an adequate supply of land to meet the future needs of strategic opn space links, natural areas and recreational facilities on any future space tourism destinations, increasing pressure for public access to open space areas with conservation values, competing interests between adjoining land use and development on public open space and its user groups, use of public open space and recreational resources for drainage purposes, raising higher space traveller hotel residential development placing increased pressure on the demand for public open space planet land use aim and developing public open space mor intensive leisure and sport activities on any future new space tourism planet destinations.

When the space tourism leisure providers have long term objectives to attempt to solve above these any one of key issues. It will ahve a more clear objective to achieve its long term space tourism leisure business market. It's long term objectives can include such as below:

To identify existing and future active and passive recreation needs and social trends of future space tourism visitors; to provide a wide range of high quality and accessible public open space public land areas to encourage physical activity and social interaction to meet the existing and future needs of space travelling visitors; to identify existing gaps in the public open space network and develop any different kinds of space trip arrangement to satisfy the different identified target space traveller individual needs; to protect enhance and increase landcrapt values of public open space land use; to recognize the hierarchy of public open space assets; equitably distributing open space resources; access to facilities and a diverse range of opportunities to incorporate the drainage function in public open space travelling destination areas without detriment to safely, environmental, visual and recreational values.

So, these development of any space planets howo to use their lands objectives will bring long term space travelling destination beneficial advantages to raise to build the space hotels, space swimming pools, space gardens, space cinemas, space sport places to let future space travelers can stay in Mars or Moon planet destinations to enjoy these leisure facilities and they can feel which are similar to our earth leisure facilities attractively.

These space buildings are important to attract future space travellers to catch spacecraft to fly to Mars or Moon planet to travel in possible because it is fun and exciting space trip when these leisure facilities can be built on Moon or Mars to let space travellers to stay short days in either these two planets to live their space hotels. So how to build any one of these space leisure building which is another important objective for any future space tourism leisure business, instead of how to arrange any space destination trip objective. So, any space tourism leisure provider ought not neglect how to achieve these two main space tourism objectives.

However, these are key questions continually asked regarding the viability of space tourism. They concern financial, marketing and political communities. Their concerns can be best addredded in a properly, comprehensive business plan. Some questions can not

be answered definitively at this time. Hoever, knowledge of the concerns and developing space businesses in any space traveling leisure planning stages and efforts to raise capital in the following questions, every spce tourism leisure business leader needs to concern this questions as below:
Can the space tourism industry into a profitable enonomic industry?
Are challenges related to financing, marketing, business methodologies or a combination of all of these facets?
Can the proponents of space tourism to be proven business tools and methodologies in their presentation of an acceptable business plan?
Can at least a cost effective, certified passenger space tourism journey to be developed for space tourism?
What effects will influence space-tourism businesses of NASA begins selling seats on the US space shuttle to civilian space tourists?
All above questions will be every new space tourism leisure businessman who needs to concern questions in order to achieve whose marketing strategy more successfully. Consequently, marketing strategy is important to be prepared in order to follow corrective steps to achieve every space tourism leisure business missions and objectives more easily.

- Space tourism leisure behavioral economic consumption model

In space tourism leisure industry, due to every time space trip needs the space travelling planner to plan how much budget to consume expensive spce ticket price. So, it seems that the target customers will be rich or high income level young people or the retirement rich old people target customer group.
So, it brings this question: How to persuade these rich or high income young people or rich retirement old people to prefer to spend spce tourism leisure at least one time in their life?
It is one valuabe research question to every future space tourism leisure provider. I shall indicate the successful factors to analyze

how to persuade them to accept this kind of potential space travelling leisure in behavioral economic personal consumption view point, in order to explain the cause and effect relationship between of these factors as below:

(1) Economic environment variable factor

Firstly, it is economic environment variable factor whether it can influence to space tourism leisure consumption changing. As I discuss about economic environment variable issue will influence consumption behavior changing. For space tourism leisure case, it is not now kind of essential consumption leisure product to every one. So , even the rich or high income people who will be influences to seek this kind of leisure to play, it the economic environment is improved, it will influence they have positive attitude and interest to choose this kind of leisure consumption. However, if the economic environment is worse, it will influence they have negative attitude and no interest to choose this kind of leisure consumption, due to space travel is one kind of expensive leisure consumption to every one.

Hence, in this space tourism leisure industry, it does not ensure that the rich or high income people must be persuade to choose this kind of expensive space tourism entertainment in whose holiday or retirement time. They can have the common tourism entertainment to go to different countries to travel many times in our earth. Otherwise, space tourism leisure is more expensive to compare common earth tourism leisure , it means that the rich or high income people only spend one time spacecraft catching to fly to space to travel in their life, it is more difficult to every space traveler like to catch spacecraft to fly to space to travel more than one time, due to he/she had attempted to catch spacecraft to fly to space to travel to own space travel experience, he/she will feel enough satisfactory and enjoyment in common. Hence, it is possible that future many rich or high income people only like to spend one time space tourism leisure, then they won't continue to spend this kind of tourism entertainment again in their life.

Thus, space tourism leisure providers need to arrange any special

or attractive space tourism leisure to persuade these high income or rich target clients to consume, when the economic environment will change worse. The Europen space agency (ESA), defines this phenomenon between economic environment variable and space tourism client growth or falling number relationship as: " space tourism is an execution of sub-orbital flight by privately finded and/ or privately operated vehicles and the technology development driven by space tourism market."
it seems that space vehicle is one attractive travelling desire tool will be one attractive selling point to influence space tourism leisure consumer individual entertainment choice or attitude to be changed to positive leisure consumption attitude to prefer to play this kind of space tourism activities when economic environment changes to worse. Hence, when economic environment is worse, the economic wore changing factor will influence the space travelling planner individual leisure consumption desire, even it will influence the rich or high income young people or rich retirement people target customer both groups.
As (ESA, 2008) indicated space vehicle will be one kind of attractive leisure tool for spce traveler. So, I suggest that space tourism lesiure journey arrangement needs to include that such as : the space travelers can catch space vehicle to move on Moon or Mars plants land to feel what the different feeling is between during they are catching public transportation tool, such as bus or taxi during the are catching these transportation tools on earth land and during they are catching space vehicle tools on Mars or Moon planet's lands. It is so exicting and fun catching space vehicle tool experience on these both Mars or Moon planets' lands to the young and old age space travelling passengers. Because every space vehicle's speed is not very fast and it will move on Moon or Mars planets slowly. So, any aged pace travelling passengers can attempt to play this kind of space facilities leisure after they catched spacecraft to fly to these both Mars or Moon planet to stay. They can spend half hour or one hour, even more than one hour to catch the space vehicle to go to anywhere on Mars or Moon to travel. It

is possible that they can find exciting and undiscovered things on these both planets.

So, catching space vehicle to go to anywhere on either these both planets journey, it will one essential part of space travelling journey during the economic environment is changed to worse. It is extra attractive space travelling leisure journey to attract space tourism consumer individual leisure desire when economic environment is worse.

Hence, from this perspective then space tourism could be understood as a section of the tourism industry mainly based on technological development, progression and its activitity being related specifically to sub orbital flights. So, if future space tourism providers expect whether the global economic environment changing will be better or worse which won't influence space tourism leisure consumption desire to be changed. The space tourism leisure providers need to persuade the space tourism planners feel space tourism would have to be treated like an already exciting part of the tourism industry. It means that space tourism leisure is one kind of tourism leisure choice to replace common earth tourism leisure consumption. When travelers feel space tourism is another tourism leisure to replace which can replace common earth tourism leisure. It will avoid the worse economic environment changing factor to reduce the rich or high income young people or rich retirement old people whose space travelling leisure consumption desire.

Consequently , the question in relation to, in what kinds of space tourism journey message do space travel providers promote behind whether space vehicle journey promotion message which is needed when economic environment will change worse. I shall be asked, as understanding the meaning in which space tourism is being marketed, communicated is seen as a factor , which can either positively contribute to future development of the tourism industry or lead into prolonging or seen stopping the space tourism industry from its progression.

(2) Space tourism leisure journey management factor

Secondly, space tourism leisure jounrey management factor, how to arrange every space tourism leisure journey which will be one important factor to influence space tourism planner individual tourism consumption desire.
In general, it can includes these several forms of space tourism leisure activities in every space lesiure trip arrangement. The following classification of space tourism include: Terrestria spce tourism (i.e. NASA visit centre, space movies, online space experience); Atmospheric space tourism (i.e. : MIG 31 flight, zero G. flights) and astro (orbita) tourism (i.e.: trips to the international space station-beyond earth orbit) (Cater 2010, Crouch et al. 2009).
Instead of US domestic space tourism market is potential, next country is Japan. First, the study is made by Collins et. al (1994, 1996) in Japan on 3030 research participants, showed that 80% of respondents under the age of 50 were willing to travel to space and out of them 20% were willing to pay year's salary for the space travel experience. Yet, it could be citicized that the Japan people age group of under 50 could be too broad, in general different generations under one groups. nest besides the willingness to go to space, the Japanese study showed respondents motivations for travelling to space, including any fun and exciting attractive space tourism journey, e.g. interest in space walk, catching space vehicle or driving space vehicle on the either Moon or Mars planets, earth view, zeo gravity experience, livin gin space hotels one night or more, watching movies in space cinemas, swimming in space pools, visiting space gardens, running in space sport centers, catching spacecrafts to view earth or Moon or Mars planets.
Hence, it seems attractive space tourism journey can persuade another country's space travelling planners, such as Japanese attempts to satisfy whose space tourism needs. So, different kinds of attractive space trip journey arrangement will be one important factor to influence young and old age travelling consumption desire.
It implies that attractive space tourism journey will be one

influential factor to encourage other countries tourism consumers attempt to another kind of leaving earth tourism leisure.

So, any space tourism trip destinations and leisure facilities arrangement must need to satisfy space traveler individual leisure needs and every space trip must be more fun, exciting and comfortable and enjoyable feeling to compare general tourism journey in earth. Due to general earth tourism leisure will be space tourism leisure's competitive or replaced leisure product and service. Hence, space trip destinations and leisure facilities choice will be one important factor to influence space travelling planner's consumption desire.

Every space travelling planner will compare general earth travelling leisure's destinations and leisure facilities arrangement whether the space travelling trip arrangement , leisure facilities arrangement and food arrangement, space vehicle or spacecraf leisure comfortable influence issues which will have more satisfactory enjoyable feeling to compare general earth tourism leisure and their spending expenditure to every space trip whether is value or is not value.

Consequently, economic environment changing factor and space trip and leisure facilities arrangement factor which both will influence any space tourism planner individual consumption desire mainly. So, space tourism businessmen ought concern these two aspects of factors how and when will change to adapt any country's potential space traveler's space tourism changing taste and needs in order to follow the new space tourism changing needs easily.

● Space tourism market moral ethic risk threats

What are space tourism moral ethic risk during the space businessmen operate this businesses as well as what market threats who will encounter to face difficulties ? I shall give actul cases to explain how and why these challenges will cause to influence any new space tourism businesses development successfully.

(1) Potential accidents aspect

Firstly, space travelers will concern that public reactions to potential accidents aspect during they are catching spacecrafts to travel to space. In fact, it is moral ethic responsibility to any space tourism leisure providers to provide safe, comfortable and non accident occurrence in their whole space trip. Because once time accident will cause any one of space passenger hurt or death. So , it must be any space tourism businessmen responsibilities to concern whether they have enough confidence to ensure none any accident occurrences in every space tourism trip.

Hence, in space tourism industry, government needs have public policy to threaten or prohibit any space tourism leisure providers neglect to often check and ensure any spacecraft machines or equipments are regular opeations, as well as often renew new spacecraft machines when they are old to be used. The policy is a force effort to need them to abide every space tourism leisure safe responsibility to ensure or guarantee any one of spacecraft won't have accident occurrences during it has left earth to fly to space in whole space trip journey from the beginning to the end till to the spacecraft come to earth safely.

Hence, this policy forces any space tourism leisure providers concern to put a monetary value on increased or reduced risk of death, the " value of statistical live", used to characterize when the benefit of safety regulation is worth the cost such regulation improves. So, the country government and the country's space tourism leisure providers both have responsibilities to guarantee all space tourism passengers' life safety. It must not allow any death or hurt occurrences during every space tourism trip.

Even, the country government can have legal action to publish any space tourism leisure providers, when their every space tourism trip has occurred accidents, e.g. fire accident occurrence in spacecarft or spacecrat machines are broken to be damaged and need to be repaired during the space tourism trip. It will threaten to reduce trip accident occurrence, such as this cases. The commercial space ventures may present risk to property as well, such as a fire starting on the ground by launch-related material or problems presented by space debris.

In principle, liability law can provide incentive to deter carelessness that could lead to the destruction of property, although statutory (rather than common law) assignments of liability for commercial launches are somewhat problematic.
Consequently, if the space tourism leisure provider expected to grow space tourism passenger number in long -term time, it must need to ensure none any accidents can occur during any space trip. Otherwise, the space tourism passengers can choose another space tourism leisure provider to replace its spce tourism leisure easily.

(2) Space tourism destinations and space tourism entertainment facilities safe arrangement challenges aspect

Secondly, it is space tourism destinations and space tourism entertainment facilities safe arrangement challenges. Nowadays, commercial space travel is looking more like a real possibility than science fiction. The usual ethical issues related to the safety of the space destination choices and the space tourism entertainment facilities, e.g. space vehicles, space hotels, space swimming pools, space sport centers, space cinemas, space gardens, space farming lands. In this strange space environment and safety concerns are just the beginning as there are othe interesting questions, such as below:
What likely would be a fair process for commercializing or claiming property in any space planets? Such as Moon or mars, when any future space tourism leisure providers who need to build above these any one of space entertainment facilities on these planets to provide to their space travelling customers to play.
How to distribute and manage these any lands ownership to these future space tourism providers fairly and legally?
How likely would a separatist movement be among space settlements to want to be free and independent states?
How to ensure above future space entertainment facilities and space entertainment places are in the safe space environment to be provided to any space travelers to play in any planets, e.g. Moon or Mars etc. planets.

So, concerning how to arrange space entertainment facilities to provide to space tourism clients to play in any safe space environment issue, it will be another concerning question to every space tourism leisure providers. When they decide to choose anywhere to the space hotels, space swimming pools, space gardens, space cinemas or space farming lands or space sport centers. These space buildings will need to be built in the safe, on stable stone lands environment and none any natural distaster, such as large wind or space underground water etc. unpredictable space natural distasterr attack to these space buildings suddenly. Because it has responsibility to any space tourism leisure providers to guarantee any one of these space buildings are safe to be built in the planet's safe land environment. It aims to achieve none any accident occurrences during their space tourism clients are staying to enter these any one of space buildings to visit or play any space entertainment facilities safely, e.g. space vehicle.

So, they must need to ceck anywhere the space planet's places to be ensured safe to build any buildings. Then, they can choose the suitable locations to build space entertainment facilities or buildings more confidently.

In fact, any space entertainment facilities, e.g. space hotels, space farming lands as well as space transportation tools, e.g. spce vehicle, spacecraft , these things will be value to be concerned to any space tourism leisure providers and it is business moral ethic responsibility to every one of them, when they plan to develop their space tourism business in any planets.

(3) Space tourism market competition challenge aspect

Thirdly, any provate space tourism development leisure businesses will face market competitive challenge, such as large spacefaring countries, e.g. US, UK have possible to dominate future space tourism leisure business (government can own space tourism leisure business). They will be main actors in space were nation-states. Large spacefaring counties can build the space vehicles, that can take people and cargo into orbit and to the Moon, or Mars crafted international space law and shaped the main investments in space tourism leisure technology.

So, it is possible that the own space technological developed countries, such as US, UK, these countries governemts will have possible to operate public fund to support space tourism leisure business. It implies that private space tourism leisure businesses will face public space tourism leisure business and themselve private space tourism leisure business market competition in space tourism leisure industry.

If these two countries governments also participate this private space tourism leisure market. It will raise market threats to any private space tourism organizations.

Whether will developed countries governments participate private space tourism market? It is possible that new commercial actors began to enter the space tourism leisure industry, looking to disrupt both space launch services ans use space in new exotic ways. For example, the US government also moved its purposeful degradatoin of the global positioning system (GPS), so US government will have effort to dominate GPS global positioning system communication business also. As this GPS communication business case, future US government has possible to decide to participate space tourism leisure business also.

However, in the future, space tourism leisure industry may contribute even more the developed countries, e.g. American, England economy. Space tourism and resource recovery, e.g. mining on planet, Moons and asteroids in particular may become large parts of that space tourism industry if these countries governments participated to this space tourism industry development. Of course, their viability rests on a range of factors, including costs , future regulation, international market competivitive problems and assumption about space technological development. However, these is increasing optimism in these areas of economic production to bring human space tourism leisure enjoyment and space mining resource development benefits. But the space economy is not just about what happens in orbits or how that alters life on the ground. The growth of this economy can also contribite to new innovations across all future possible unpredictable or undiscovered technological development, instead of space tourism leisure or space mining resource exploitation development.

Consequently, any space development technological governments will have possible to bring economic benefits from either only private space tourism leisure organizations or governments and private space tourism leisure both organizations cooperate to participate to achieve space tourism misson to contribute to global economic development and create new jobs to be employed in space labor supply market. The most important successful factor is that space tourism entertainment companies need to learn how to apply (AI) technology to assist any space boats and space journeys and any space entertainment facilities in order to raise safe and efficient service performance to let space passengers to feel satisfactory and confident in every space tourism (AI) design journey.

Reference

Cater Iain , Carl 2010, " Steps to space: Opportunities for astro tourism development, tourism management 31 (2010); pp. 838-845; Elsevier Ltd, DOI: 10:1016/j.tourman. 2009.09.001

Collins Patric, Iwasaki Yoichi, Kanayama Hideki, Ohnuki Misuzo 1994, comercial implications of market research on space tourism. journal space technology and sciences , vol. 10 no 2, 94 Autumn, pp.3-11. copyright: Japanese rocket society; available at: www.spacefuture.com/archive/commercial-implications-of market-research-on-space- tourism.shtml.

Collins Patric, Marita M; Stockmans R. and Kobayahi S. 1996. "Demand for space tourism in America and Japan and its implications for future space activities ". sixth international space conference of Pacific basic societies; Marina del rey; California: Advantages in the Astronautica science (AAS paper no AAS 95-605) vol. 91. pp. 601-610. Available at:
http://m.internationalaerospaceconsulting.org/upload/space % 20Future%20-%20Demand% 20for%20space% 20Tourism%20in%

20America%20Japan.pdf

ESA 2008, " Richard Garriott, millionaire American space tourist. blasks off of international space station". published in 12.11.2008. Huffington post, seen on i01.04.2015; available at: http://www.huffington.com/2008/10/12/richard-garriott-milliona-n-1333940.html.

Klemm, G., & Markkanen, S. (2011). IN A Papathanassis (ed.) The long Tai , tourism (pp.95-103). Weisbaden, Germany : Gabler Verlag; Springer Fachmedien Weiesbaden GmbH.

KSCVC Visitors, 2013; MRI 2013 Market by Market

Von Der Dunk , F. (2012). The integrated approach. Regulating private human spaceflight as space activity, aircraft operation, and high-risk adventure tourism. Acta Astronautica, 92(2), 199-208.

Chapter 8 AI brings positive or negative impact to our societies

The relationship between robotic invention and economic growth

Can robotic invention bring social economic growth? If it is possible that why and how robotic invention can assist any countries social economic growth? First, we need to know whether what economic growth means in order to answer this question. In macro economic view, economic growth may mean that GDP growth, employment ratio growth, job creating growth, unemployment ratio reduces, consumption growth, productive industries growth, service provision and service needs increases. So, it seems that any countries' social development , when it can have positive impact growth for any one of these issue. It implies that the country's economic is growing.

To discuss AI and economic growth issue, it may being these two main questions. Whether robotic invention may have direct or indirect relationship to influence any countries' economic growth? What are the main factors to cause robotics invention to bring the country's economic growth in its society?

On the first hand, some researchers believe that robotic invention may influence any countries' economic growth. However, otherwise, other many researchers find large and robust negative effects robots on employment and wages. They estimate what are more robot per thousand workers reduces the employment -to- population rate by between 0.18 and 0.34 percentage points, and is associated with a wage decline of between 0.25 and 0.5 percentage. Because many jobs can be replaced by robotics. So, it seems that robotics invention can reduce workers number and increasing wage level, even robotic increasing number, it can influence unemployment ratio increases to the country. But, some lecturers estimate that it can impact economic growth. They estimate that AI may deliver an additional economic output of around US$13 trillion by 2030 year, increasing global GDP by about 1.2% annually. This will mainly come from substitution of labour by automation and increased innovation in products and services.

What is the impact of robots on society? They may include this spillover, one robot per thousand workers has slightly less of an impact on the population as a whole, leading to an overall 0.2% point reduction in the employment-to-population ratio, and reducing wages by 0.42%. Thus, adding one robot reduces employment nationwide by 3.3 workers. So, it seems that robotic number increases may influence global workers number increases many influence global workers number reduces as the same time. It can cause unemployment workers number increases in societies.

On the other side, robotic invention can increase productivity, because robots increase productivity, which means that fewer human hours are needed to produce a given output. But, higher productivity also reduces production costs and output prices. Consequently, robotic production anticipation , it can increase the quality demanded by consumers, and firms hire workers in this increased demand.

But, when robotic production participation to any industrial manufacture, it can also bring negating effects, when they are entering the workforce, e.g. higher maintenance and installation costs to factories, enhanced risk of data breach and other cybersecurity issues, reduced flexibility, anxiety and insecurity regarding the future social development. So, it seems that robotic invention can bring the future of workplace automation may reduce workers number of factories, loss of jobs and reduced employment opportunities to global future workers, potential job lose, initial investment costs to employers. However, AI may also bring harmful to our future societies because if AI surpasses humanity in general intelligence and becomes " superintelligent" , then it could become difficult or impossible for humans to control. Moreover, a second source concern is that a sudden and unexpected " intelligence explosion" might take an unprepared human race by surprise. Hence, robotic invention may become human enemy or soldier, if human applies it to social damage aspect.

However, robotic invention can also bring advantages to our future societies in possible, robotic automation may bring advantages to employers : cost effectiveness, improved quality assurance, increased productivity, avoiding workers need to work in hazardous environments. But,. AI can also bring positive impact our daily lives, such as artificial intelligence can dramatically improve the efficiencies of our workplace and can argument the work humans can do. When AI takes over repetitive or dangerous tasks, it frees up the human workforce to do work, they are better equipped for, tasks that involve creativity and empathy among others.

However, robotic invention may bring organizational benefits in our societies. Robots will have a profound effect on the workplace of the future. They will become capable of taking on multiple roles in organizations, e.g bookkeeping or writing law draft simple clerical tasks. So, it's time for us to start thinking about the way we shall interact with our new coworkers. To be more precise, robots are expected to take over half of all low-skilled jobs in our societies, e.g. cleaning , warehouse picking up delivering tasks, restaurant cooking, hotel front line customer service, hotel room food delivery tasks etc.

So, when robots would be used in many fields all over the world. However, robots can not totally rule over the workplace by replacing all humans at jobs to keep economy afloat. Hence, robots in the workplace may bring advantages, they will not have bad emotion problems, such as safety of utilizing robotics to work in dangerous workplace. Robots do not get distracted or need to take breaks robots never need to sleep or they need to divide their attention between a multitude of things, perfection, let employees to feel happier and safe to work when robotics can help workers to work in dangerous warehouses or factories, workers can be replaced from robotics, increases productivities and job creation , even raises efficiencies in any workplaces.

In our future, whether robots won't destroy humans. It depends on how humans choose to apply this technological worker tool. A robot may not injure a human being or, though inaction, allow a human bring to come to harm. A robot must obey the orders given it by human beings, expect where such orders would conflict with the first law. A robot must protect its existence as long as such protection does not conflict with the first or second laws. How AI technology affects us in the future. They are concerned that we will see increases in stress, anxiety, and depression as digital lives expand. Meanwhile , we shall need to adapt our future digital living, there will be less face-to-face interaction , increased inactivity, poor in-person communication skills and an overall distrust among people. For robotic invention case example, future robotics can may focus on developing these five major field: Human -robotic interface, mobility, manipulation, programming, sensors and their importance to robotics development.

Robots can be applied to educational aspect. Robots can be used to bring students into the classroom that otherwise might not be able to attend. Robots such as the one mentioned are able to bring school to student who can not present physically. Simulators-high school sees the strongest example of stimulators within drivers' education courses, e.g. Google's worker robots. Google is planning to produce worker robots with personalities. So, robots can help teachers automate key classroom processes, integrate advanced technologies, acclimate students to technological change and helping identify personalized learning potential and discovering key learning trends.

- How manufacturers to raise efficiency in order to improve economic growth in possible to apply manufacturing robots to improve help economic growth

(AI) can be defined as the capability of a machine to imitate intelligent human behavior. If AI can imitate any talent humans to learn how to improve their behaviors, e.g. manufacturing industry worker behavior improves to raise GDP growth or productive number grows rapidly . Can artificial intelligence being labour to become automated? Allows an ever-in-to increasing number of tasks previously performed by human labour to become automated in the ordinary production of goods and services process.

Can (AI) create new ideas and technologies to help businessmen to solve complex problems and improve automation in the production of goods and services. The question concerns how (AI) manufacturing technology can impact economic growth?

(AI)'s new form of automation live, self-driving cars, or they may bring high levels of skill,such as legal services, radiology, and some forms of scientific lab-based research. Can it allow our societies be impacted on economic growth, due to automation to disicipline our modeling of AI.

In fact, AI automation to production of new ideas to future any industrial manufacturing process. It can influence to market structure, organization restructure, reallocation and wage inequality. So, discovering to AI automation can help future any organizations need to change existing task or discovering new tasks that can be used in production a reflects the fraction of tasks that have been automated.

It brings automating old tasks when automated could be constant, leading a stable , capital share and a stable growth rate. Hence, in long term, AI automation manufacturing technology may help businessmen to reduce large manufacturing cost in order to achieve stable micro economic benefit to them when they can apply AI automation manufacturing technologies to improve their productivities in efficient manufacturing method. For Coca Cora soft drink example, if Coca Cora applies AI automation to raise its productive soft drinks number. Then , it can manufacture many soft drinks number in short time per day. Then, its sale number can be increased, when it has enough number to supply to global to sell its soft drink. Consequently, soft drink sale number must grow rapidly significantly.

The fact that automated goods are produced with cheap capital , but it can also help business to raise production number significantly . How this superintelligence affect the economy? It seems physical tasks are essential to producing output, but when the manufacturer applies robotics to help production . Then, employees number may reduce, due to robotic participation, wages expenditure may reduce , but production number may increase .

So, AI increases the motivation at physical tasks. Hence, AI must may bring production growth innovation incentives to any manufacturers. Finally, with imitation and learning being performed mainly by super machine in developed economies. Then , research labor would become devoted to product innovations increasing product variety or inventing new products (new product lines), to replace existing products.

It is one good example to explain that how AI can bring long term social economic benefit to any manufacturers, when any old (existing) product lines are improved to innovative new product line by robotic manufacturing participation. Moreover, AI can change market structure to be reduced competition. When be escape competition effect tends to dominate at low discouragement effect may dominate for higher levels of competition or in less advanced economies. Hence , AI can also affect innovation and growth through potential effects, it might have on product market competition. So, it seems that (AI) can respond on helping social economic growth principle.

On conclusion, although robotic invention may bring disadvantages to raise unemployment ratio in possible when there are many low-skilled jobs are replaced by robotics, but at the same time, robotic may be applied to education aspect, when we are experiencing digital knowledge social development stage. Hence, smart robots ought may help humans to raise economic long term growth in possible when robotics can help the developing countries to develop more rapid to be developed countries as well as they can help the developed countries to develop more advanced both in global whole one economic developed societies. So, when robots can assist any countries to cooperate together, any countries are not independent, we need robots to assist develop between countries. Then, robots can assist any countries to develop economic growth in possible in future this day comes.

How robotic helps to solve recession

Is the use of robots to job market increasing during the Great Recession?

Some researchers constructed a measure of the use of robots—commonly referred to as "robot intensity"—to estimate trends in robot exposure across more than 250 metropolitan areas and over time, finding that: During the Great Recession, robot intensity plummeted. But since 2009, robot intensity has sharply increased nationwide. They felt that robots may influence recession in possible. How are robots going to affect our jobs? Most analysis tends to be prospective in nature, and estimates of future impacts on employment vary widely, with some studies predicting that as many as 50 percent of all workers are at risk of losing their jobs to automation. Even less is understood about the actual impacts of robots on jobs, wages, and workers today. If there are many low skillful jobs e.g. cleaner, warehouse deliver, cooker, restaurant waitor etc., even high skillful jobs, e.g. accountant, lawyer, doctors etc. occupations are replaced by robotics. Then, our societies will increase unemployed people number. Consequently, great recession will be caused by robotic workers because when our societies have many people lose jobs, then many people loss income, then our consumption desires may be influenced to reduce. Consequently, many businesses may lose many customers. Low consumption desires may bring serious recession to any countries, due to robotic workers number increases to replace human workers in global societies.

The reason is that new technologies of the period have enabled people to be very productive while working part-time. Businesses do not need large numbers of employees, so individuals can devote most of their waking hours to hobbies, volunteering, and community service. In conjunction with periodic work stints, they have time to pursue new skills and personal identities that are independent of their jobs. Developed countries may be on the verge of a similar transition. Robotics and machine learning have improved productivity and enhanced the economies of many nations. Artificial intelligence (AI) has advanced into finance, transportation, defense, and energy management. The internet of things (IoT) is facilitated by high-speed networks and remote sensors to connect people and businesses. In all of this, there is a possibility of a new robotic society that could improve the lives of many people, but it also encourage future businesses apply robotics to replace human workers to do many jobs in finance, transportation, defense and energy management, medical health, hotel, tourism, cinema, theatre etc. entertainment service fields. So, robotics may bring reducing cost benefits to businessmen, but they can also increase workers losing jobs number in our societies if future many businesses make decision to apply robotics to replace human workers to do any simple ot complex jobs in global job market.

A McKinsey Global Institute analysis of 750 jobs concluded that "45% of paid activities could be automated using 'currently demonstrated technologies' and . . . 60% of occupations could have 30% or more of their processes automated."[6] A more recent McKinsey report, "Jobs Lost, Jobs Gained," found that 30 percent of "work activities" could be automated by 2030 and up to 375 million workers worldwide could be affected by emerging technologies.

Researchers at the Organization for Economic Cooperation and Development (OECD) focused on "tasks" as opposed to "jobs" and found fewer job losses. Using task-related data from 32 OECD countries, they estimated that 14 percent of jobs are highly automatable and another 32 have a significant risk of automation. Although their job loss estimates are below those of other experts, they concluded that "low qualified workers are likely to bear the brunt of the adjustment costs as the automatibility of their jobs is higher compared to highly qualified workers."

reference

James Manyika, Susan Lund, Michael Chui, Macques Bughin, Jonathan Woetzel, Parul Batra, Ryan Ko, and Saurabh Sanghui, "Jobs Lost, Jobs Gained: Workforce Transitions in a Time of Automation," McKinsey Global Institute, December, 2017.

Melanie Arntz, Terry Gregory, and Ulrich Zierahn, "The Risk of Automation for Jobs in OECD Countries," Organization for Economic Cooperation and Development, Working Paper 189, 2016.

However, some economists felt opposite opinions, they believes that future many businesses won't choose whole applying robotics to replace human workers in global job market. So, they are only human workers assistant role. Economists have, on the whole, been fairly discuss about the impact of robots and AI on workers. History is strewn with incorrect predictions of the looming irrelevance of human labour. The economic statistics have yet to signal the arrival of a robot-powered job apocalypse. Outside of slumps, firms remain keen to hire humans, for example. Growth in productivity—which ought to be surging if machines are helping fewer workers produce more output—has been unimpressive. A look beneath the aggregate numbers, though, reveals that change is indeed afoot. They believes that an AI-induced change in the mix of jobs need not translate into less hiring overall. If new technologies largely assist current workers or boost productivity by enough to spark expansion, then more AI might well go hand-in-hand with more employment. This does not appear to be happening. Instead the authors find that firms with more AI-vulnerable jobs have done much less hiring on net; that was especially the case in 2014-18, when AI-related vacancies in the database surged. But the relationship between greater use of AI and reduced hiring that is present at the firm level does not show up in aggregate data, the authors note. Machines are not yet depressing labour demand across the economy as a whole. As machines become cleverer, however, that could change.

Take work by Daron Acemoglu and David Autor of the Massachusetts Institute of Technology, Jonathon Hazell of Princeton University and Pascual Restrepo of Boston University, which was presented at the recent meeting of the American Economic Association (AEA). The authors use rich data provided by Burning Glass Technologies, a software company that maintains and analyses fine-grained job information gleaned from 40,000 firms. They identify tasks and jobs in the dataset that could be done by AI today (and are therefore vulnerable to displacement). Unsurprisingly, the researchers find that businesses that are well-suited to the adoption of AI are indeed hiring people with AI expertise. Since 2010 there has been substantial growth in the number of AI-related job vacancies advertised by firms with lots of AI-vulnerable jobs. At the same time, there has been a sharp decline in these firms' demand for capabilities that compete with those of existing AI.

An AI-induced change in the mix of jobs need not translate into less hiring overall. If new technologies largely assist current workers or boost productivity by enough to spark expansion, then more AI might well go hand-in-hand with more employment. This does not appear to be happening. Instead the authors find that firms with more AI-vulnerable jobs have done much less hiring on net; that was especially the case in 2014-18, when AI-related vacancies in the database surged. But the relationship between greater use of AI and reduced hiring that is present at the firm level does not show up in aggregate data, the authors note. Machines are not yet depressing labour demand across the economy as a whole. As machines become cleverer, however, that could change.

Evidence that AI affects labour markets primarily by taking over human tasks is at odds with some earlier studies of how firms use the technology. A paper from 2019 by Timothy Bresnahan of Stanford University argues that the most valuable applications of

AI have nothing to do with displacing humans. Rather, they are examples of "capital deepening", or the accumulation of more and better capital per worker, in very specific contexts, such as the matching algorithms used by Amazon and Google to offer better product recommendations and ads to users. To the extent that AI leads to disruption, it is at a "system level", says Mr Bresnahan—as Amazon's sales displace those of other firms, say.

New work by Ajay Agrawal, Joshua Gans and Avi Goldfarb of the University of Toronto suggests that this state of affairs may not persist for long, though. As the quality of AI predictions improves, they write, it becomes increasingly attractive for AI-using firms to restructure in more radical ways. At some level of accuracy, for example, Amazon's ability to predict consumers' desires could encourage the firm to adjust its business model—by pre-emptively shipping goods to consumers before they ever go searching at Amazon in the first place—in ways that are likely to change how many workers and of what sort the firm requires. In that event, the influence of AI on the economy could change dramatically. So, such as Amazon case, it applies robotics are only concentrated on predicting consumer prediction aspect, AI is only assistant role to Amazon human market researchers. They help Amazon market researchers to gather consumer behavior data , but Amazon human market researchers need to do marketing analysis tasks by themselves. So, Amazon can not employ human market reseachers jobs position in itself company. Amazon needs robotic and human market researchers to do market research tasks in order to achieve how to predict consumer behavior more accurately. Thus, it seems that future large enterprises won't fire any professional staffs more easily because some complex tasks, e.g. analysis tasks, they believe that human's analysis can make more accurate judgement to compare AI's analysis.

However, some scientists believe that some skilful professional occupations , they have possible be replaced by robotic. Will Architects and Engineers be Replaced by Robots? It's not uncommon for people to think they may be replaced by a robot in the workplace. After all, it's happened plenty of times before. For example, the rise of the mechanical assembly line saw machines replace people in the early 20th century. With recent advances in artificial intelligence (A.I.), it's entirely possible that more jobs are at risk. Even skilled workers, such as architects, programmers and engineers may be at risk. One day, an A.I. software developer may be able to do everything that a human programmer can do.

Recent reports have not abated this thought process. In fact, the 2016 Economic Report of the President seemed to suggest that artificial intelligence is playing an increasingly important role in the engineering industry. Just think about the software you use in your work. Many software packages can handle a lot of the complex calculations for you. Yes, this cuts down on the amount of work you do. However, this automation may also present a threat to your job. What if the future sees these same software packages handling data input, as well as processing.

Automation is important. The use of artificial intelligence, alongside various other technologies, has always improved production. More work gets done, which means that businesses make more money. Architects and engineers constantly look for ways to speed up their work. The desire for automation has informed many recent software innovations. Furthermore, project methodologies, like Building Information Modelling, place automation at the fore. That's great for speed and efficiency, but what does it mean for architects and engineers? History has shown that automation has a very human effect. People lose their jobs because machines can do them faster. Just think of it from a business viewpoint. Do you want to pay 10 or more employees, or invest in one machine? More often than not, the machine will cost less than the employees, even if you factor maintenance into the equation. It's a simplification, but not an invalid one. Businesses make these sorts of decisions all the time. By pushing for automation, architects and engineers may be slowly working themselves out of their own jobs.

Several studies have also suggested that artificial intelligence may cause job losses. One recent example comes from the University of Oxford. The study found that over 700 types of jobs are at risk of technological disruption. All told, this means that about 47% percent of jobs are at risk because of artificial intelligence. That is a huge amount of people who may find themselves obsolete due to advancing technology. The same study also mentioned a concept called the "technological bottleneck". The researchers used this to determine how "at risk" a job was of displacement. The bottleneck takes three factors into account:

·How much creative intelligence the role needs

·If manual manipulation and perception is required

·The role of social intelligence in the role

If a role requires a high degree of any of those three things, it's less likely that it's at risk from artificial intelligence. Architects and engineers are a good example. These professionals require a great deal of creative intelligence. Artificial intelligence and robots may not be able to emulate that creative intelligence. As a result, it's unlikely that architects and engineers need to worry about losing their jobs. Right now, at least. The study concluded with a cautionary note. It said that just because automation enhances an architect and engineer's work right now, it doesn't mean that automation won't replace that role in the future. So, if future human architects or engineers jobs can be replaced to do by robotics. Then, robotic architects can help the architectural firms to design more attractive architectural plans to satisfy construction firms clients needs in short time or robotic engineers can help the engineering design firms to design more attractive machines to satisfy any engineeing customers needs in short time , when robotics can be invented to own excellent creative ability to compare human architects or engineers. So, these two professional occupations will be lost when robotic architects and robotic engineers can be invented to own excellent creative ability to compare human architects or engineer in future one day in possible.

However, robotic architects or engineers may bring advantages and

disadvantages both aspects:
On advantages aspect:
Artificial intelligence allows us to do all of the following:
·The completion of mundane tasks that would otherwise take a lot of labour hours. Automating such tasks frees up skilled workers to work on more important tasks.
·A.I. is not as prone to making errors as a person. As long as the A.I.'s programming is good enough, you should find that calculating errors and similar issues become problems of the past.
·Speed is a key feature of artificial intelligence. Huge datasets no longer provide any problems to businesses, as automation allows for much faster processing. This means that a business can spend money elsewhere.
·The most complex A.I.s reduce the amount of risk attached to the decision-making process. The "Curiosity" Mars rover is a good example. It's programmed to choose the best course of action depending on its position.
On disadvantages aspect:
It's not all good, unfortunately. The following are some of the bad points of artificial intelligence:
·The previously mentioned job losses can cause all sorts of problems for staff morale.
·Some believe that artificial intelligence gets rid of the human element. The nightmare scenarios in films like "The Terminator" may seem far-flung, but that doesn't mean there isn't a risk in letting machines make all the decisions.
·A.I. relies on pre-existing knowledge, which means it lacks creativity. Attempting to use it for creative endeavours may result in failure.
·Algorithms may not be able to make judgement calls in disaster situations. Again, the A.I. may not take the human element into account, no matter what's actually happening on the ground.
Oe people management aspect, what do you think would be the reaction to a robot attempting to manage people? It's likely that a lot of people won't take to kindly to artificial intelligence telling

them what to do. Many underestimate the importance of people skills in the architecture and engineering profession. Architects and engineers must be able to organise workloads and manage individuals. It is sure, A.I. device could handle the former. Scheduling is a task that many already automate. However, A.I. will fall down when it comes to the human relationships that are so vital in a team environment. An A.I. won't understand when somebody is demotivated, or why. It won't make allowances for the human issues that affect every problem. This makes skilled team members even more valuable. As A.I. takes an increasing role in the workplace, the need for people management will become more important. Architects and engineers with those skills may even find they make more money to employ them. Although, it is possible that AI can replace human architects or engineers to do their tasks to be better , it can help any one architectural or enginering firms to improve design performace in order to satisfy customers design demand, but our societies will increase unemployment ratio to engineers and architects number, even universities will reduce architect and engineering students number. Our traditional professional knowledge will be felt to be rubblish when these professional subjects won't be useful to help us to find jobs easily. So, AI invention will influence many students won't choose to study these two subjects. Our societies will be influence to experience knowledge recession when knowledge will become rubblish because robotics invention , it can do many human professional jobs to do. " knowledge recession" will be important factor to bring economic recession, because human can not be encouraged to learn any new knowledge to prepare to enter job market due to robotic invention can replace us to do more complex tasks in our future societies.

Can robotic leadership be good solution method when recession had come to the country?

On leadership management aspect, can robots become clever leadership to any organizations? As artificial intelligence becomes further embedded into our everyday working lives, we are already seeing the footprint of machine learning, automation, algorithms and robots in many of our professions and sectors. However, when we look at the upper levels of business management and leadership, these technological shifts are less evident, with C level Executives continuing to lead and strategise as they have done before. In Ireland, there are more than 500 CEOs. The question is, when will we start to see machines and robots play a more central role in the CEO sphere, and is a 'Robot CEO' realistic in the short to medium term?

New research shows that 24% of people aged 25-29 would replace their boss with a robot, demonstrating an interesting trend among Generation Z. However, these data sets are perhaps less founded in AI and robotics and more in current employee engagement. It's telling that the 20-30% of people who would willingly replace their human boss with a robot is about the same percentage of people who are consistently classified as "actively disengaged" at work. In addition, research from analytics giant Gallup demonstrates that 70% of how we feel about work, namely our emotional commitment, is driven by who our manager is, again underlining the centrality of human behavioural traits when making decisions on leadership.

As the Irish economy moves forward, values will define how we use and leverage the potential of AI. Tomo Noda of the Harvard Business Review believes that we will need more focus on leadership with humanity, ethics and integrity, stating "only good people can create good AI."With many roadblocks and challenges for the economy looming, primarily in the shape of Brexit and trade tariffs, it is a sound integration of both human and tech which will provide the leadership required to ensure our economy remains robust. Human leaders have played a central role in helping to steer us out of the 2008 recession, and with diplomacy and relationship building key to our post-Brexit future, humans will undoubtedly be the key influencers within the C level for decades to come. Hence, it seems that future organizations ought choose to apply robots to assist leaders to do make important decision, if robotics can assist leaders to make any important decision to conclude the best results to improve any companies performance. Then, GDP may be influenced to increase or grow rapidly, when recession had come to the country. So, robotic leadership may be one solution to solve recession method in possible.

The Recession Cometh and Robots are Ready

In economic demand vs. supply theory indicates that consumer appetites for customized product and their expectations for ever-lowering costs. So the current tug-of-war over if, when, and where a recession will hit is not unchartered territory. For manufacturers, though, the uncertainty is particularly challenging, as the flexibility that allows operations to reflect the pace of the economy simply isn't there. The economy has been growing. Unemployment is down. Last year's Christmas sales were better than they've been in a long time. All good and logical reasons for manufacturers to hire.

Recently though, there have been signs that instability is coming. The US stock market experienced extreme volatility as 2018 came to a close. The Federal Reserve raised interest rates and laid down some pretty clear language that more was to come. Consumer confidence fell. In the UK, a deal on Brexit that would allow British manufacturers to continue to do business with the EU seemed elusive at best. The Chinese government announced that growth in its economy has slowed. And the "R" word started to appear with more frequency. These are not signs that inspire confidence. So, manufacturers once again find themselves in a place they know so well. The rock: the need to hire workers to keep ahead of demand. Compounding this challenge is that unemployment is low and it is very hard to find people with the skills needed to take a job in manufacturing and be ready to work on day one. The hard place: overwhelmingly, today's automation is fixed, expensive, and able to perform only a single task.

As one supply chain executive of a global automotive firm shared recently, "In a downturn...it is about flexibility. All of the automation we have cost too much and it is too complicated to change what it does. What we need is flexible automation that can respond when and how we need it to."Can robotics be applied to manufacturing industry to avoid cost reduces to manufacturers when consumption number reduces or recession is coming? So what makes the most sense? Hire, hoping that if and when recession comes, it will be short-lived and you won't have to lay folks off? Or try to invest in reconfiguring existing automation?

Hence, cobots give manufacturers the flexibility they need to thrive in good times and not-so-good times. Advances in robotic technology make it possible to put cobots to work

•at lower costs

•on more than a single task

•in the same amount of time, it takes to train a person – or even less

With collaborative robots, manufacturers can build the operations they need to compete and thrive regardless of the economic climate, where manufacturing robotic participation can help organizations to reduce labours number on strategic tasks and flexibility is part of the organizational human resource cost reducing strategy. It seems that robotic manufacturing workers can help organizations to reduce manufacturing cost when recession is coming. Consequently, these applying manufacturing robotic businesses may prolong business life time in possible. So, it seems that manufacturing robots may help organizations to reduce manufacturing cost to keep life when recession is coming.

VIII

www.ingramcontent.com/pod-product-compliance
Ingram Content Group UK Ltd.
Pitfield, Milton Keynes, MK11 3LW, UK
UKHW041843190726
13854UKWH00002B/701

9 798887 333243